Tokyo Listening

Lorraine Plourde

TOKYO LISTENING

Sound and Sense in a Contemporary City

Wesleyan University Press Middletown, Connecticut

Wesleyan University Press
Middletown, CT 06459
www.wesleyan.edu/wespress

Manufactured in the United States of America
Designed by Mindy Basinger Hill
Typeset in Minion Pro

We gratefully acknowledge the support of the AMS 75 PAYS
Endowment of the American Musicological Society,
funded in part by the National Endowment for the Humanities
and the Andrew W. Mellow Foundation.

Library of Congress Cataloging-in-Publication Data
available upon request

Hardcover ISBN: 978-0-8195-7883-9
Paperback ISBN: 978-0-8195-7884-6
Ebook ISBN: 978-0-8195-7885-3

5 4 3 2 1

FOR NANCY AND VIOLET, *with love*

CONTENTS

ACKNOWLEDGMENTS

This book would not have been possible without the generous participation of the numerous musicians, listeners, music writers, café owners and patrons, and background music company employees. In particular, I would like to thank the owners of music cafés Violon, Mignon, Baroque, and Marguerite, for speaking to me at length about their café's history and listening philosophies. I am deeply thankful to the employees of Japan's sound design and background music companies, who were incredibly generous with their time in allowing me to conduct extensive interviews in person and follow-up interviews over email, and for inviting me to a tour of their background music sponsored spaces in Tokyo. In particular, I would like to thank Ueki Mio at Mood Media, who shares my love of Japanese muzak. I am happy that we have remained in touch. Thanks to friends and research interlocutors in Japan: Akiyama "Cap" Tetuzi, Rie Baechel, Noel Callan, Norio Fukuda of Sweet Dreams Press, Vincent Guilbert, Haino Keiji, Kishino Yuichi, Imai Kazuo, Mike Kubeck of Super Deluxe, Kuramochi Masaharu of Uplink Factory, Itō Atsuhiro, Madoka Kono, Matsukage Hiroyuki, Mari Nitta, Odajima Hitoshi, Otomo Yoshihide, Sachiko M, Phil Snider, the late Soejima Teruto, Sugimoto Taku, Suzuki Yoshiyuki, Taguchi Fumihito of Enban, Tsunoda Toshiya, Ujino Muneteru, Unami Taku, Watanabe Hiroshi, Yonemoto Atsushi, and the staff and fellow former students at Bigakkō.

I am thankful for the friendship and intellectual support of Vicki Brennan, Andy Brown, Ryan Chaney, Michael Fisch, Marvin Gruszka, Jennifer Hale, Eric Han, Brian Harmon, Anne Ishii, Joe Hankins, Tomoko McCormick, Lauren Meeker, Jun Mizukawa, Emer O'Dwyer, Marina Peterson, Steve Ridgely, Joe Schloss, and Ayako Takamori. My Japanese skills were strengthened at the Inter-University Center for Japanese Language Studies in Yokohama in the year before beginning my fieldwork in Tokyo. Thank you to the excellent teachers, and special thanks to my brilliant teacher and friend Kushida Kiyomi, who has devoted so much of her time (both during the program and afterward) to working with me on difficult Japanese texts. I would also like to thank my fellow students and close confidants, especially Kendall Heitzman, David Offenbach-Abbott, and David Wolitz. I am lucky to have such supportive colleagues and students at Purchase College, SUNY. In particular, I would like to thank my colleagues and friends in anthropology and media studies: Rudi Gaudio, Shaka McGlotten, and Jason Pine. I am grateful for support from Linda Bastone, Darcy Gervasio, Paula Halperin, Suzanne Kessler, Mary Kosut, Michelle Stewart, Ragnhild Utheim, and Agustin Zarzosa. My friend and colleague Hakan Topal graciously provided the intriguing cover image taken inside Tokyo's Roppongi Hills Mori Tower.

The following people have given critical feedback at various stages of research, from which the book has greatly benefited: Marié Abe, Andrea Arai, Patrick Galbraith, David Grubbs, Ryan Holmberg, David J. Kim, Tom Looser, Gabi Lukacs, Noriko Manabe, Jennifer Milioto Matsue, Chris Nelson, David Novak, Shunsuke Nozawa, Jason Pine, Paul Roquet, Ethan Spielman, Carolyn Stevens, Akiko Takeyama, and Tak Watanabe. Gretchen Bakke, Scott Webel, and Matthew Wyman-McCarthy provided invaluable editing support at critical stages of the project, and I am grateful for their astute feedback. Thanks to the editor-in-chief at Wesleyan University Press, Suzanna Tamminen, and the Music/Culture series editors, Deborah Wong, Sherrie Tucker, and Jeremy Wallach, for their enthusiastic support of my book project. Thanks also to the production staff at Wesleyan University Press, including Mary Garrett and especially Sara Evangelos, whose sharp copyediting skills have greatly improved the overall manuscript. I am especially grateful to the two anonymous reviewers who offered valuable and supportive feedback in an incredibly speedy manner.

The research for this book first began during my dissertation fieldwork in Tokyo (2003–2006) and continued on numerous follow-up summer research trips from 2007 to 2016. Research was generously supported by the Fulbright-Hays Doctoral Dissertation Research Abroad Fellowship (sadly, now defunct), Social Science Research Council International Dissertation Research Fellowship (SSRC-IDRF), Weatherhead East Asian Institute, Columbia University, Shoyu Club, Northeast Asia Council for the Association for Asian Studies. I am grateful to have received the following financial support over the past nine years from my current institution, Purchase College, SUNY: Junior Faculty Development Leave Award, Peter and Bette Fishbein Junior Faculty Research Award in the Liberal Arts, Faculty Support Award, and Joint Labor-Management Individual

Development Award (United University Professions). I have presented sections of this work at various anthropology and ethnomusicology conferences over the years. In particular, I would like to thank Naomi Waltham-Smith and Michael Gallope, the organizers of an excellent panel on the political economy of sound at the American Comparative Literature Association annual meeting, where I received valuable feedback. My many music teachers over the years have left an enormous impression on my musical and sonic philosophies even if their influence is not immediately visible on the following pages. In particular, I wish to thank Lois Wynn, Jeff Fraker, Allyn Miner, and David Reck. Thanks to my advisors in Japan: Konuma Jun'ichi, who graciously assisted with my affiliation at Waseda University, and my primary advisor Shuhei Hosokawa. I benefited (and continue to benefit) enormously from Hosokawa's vast knowledge of music and sound in Japan. I am indebted to all of my graduate school mentors in ethnomusicology and anthropology. In particular, I would like to thank Ann Anagnost, Kim Brandt, Miyako Inoue, Andrew F. Jones, John Pemberton, and Mick Taussig. They gave me intellectual support, encouragement, and feedback, without which this project would never have taken shape. I want to thank my brilliant mentor Marilyn Ivy for her unflagging support of my dissertation project and for always pushing me to think critically about modernity and the very notion of the avant-garde. Although this project has transformed greatly from the original dissertation, I hope she is pleased with the final version. Her influence cannot be overstated.

A version of Chapter 1 was published in the 2014 Routledge volume *Sound, Space, and Sociality in Modern Japan*, edited by Joseph Hankins and Carolyn Stevens. Excerpts from Chapter 1 were previously published in an earlier form in the journal *Ethnomusicology* (in "Disciplined Listening in Tokyo: *Onkyō* and Non-Intentional Sounds," vol. 52, no. 2). An earlier version of Chapter 3 was published in the journal *The Senses and Society* (in "Sonic Air-Conditioning: Muzak as Affect Management for Office Workers in Japan," vol. 12, no 1). Thanks to Routledge, Taylor & Francis and the Society for Ethnomusicology for allowing me to publish these updated versions.

Last, but most important, I must thank close friends and family who have provided crucial emotional support, especially during the past few years. Thanks to Kyle Davidson, Anne Gustafson, Frances McMillen, Jennifer Sime, and Alex Wesson. Special thanks to Marie Amato, whose calm and insightful demeanor has provided critical sustenance during the final stages of this book project (thank you also for your attentiveness to my background music preferences). My immediate and extended family has been incredibly supportive and nurturing throughout the length of this project over the past ten years, especially during recent difficult times as I struggled to complete the book. Thanks to my brother Britton, sister-in-law Alina, and my amazing nephews Xavier and Leo. Thanks also to Jonathan and Kunai, TD and Linda, Rob AndristPlourde and family, Jim and Allison Plourde, and the rest of my extended family. My partner David J. Kim has been with me since this project first began in the early 2000s. His love and support, both intellectual and emotional, have sustained and influenced this project more than he will ever know.

Finally, I want to thank my parents, Nancy and René, for their endless support and love. This book simply would not have been possible without them. I feel incredibly lucky that they have always encouraged my educational and artistic pursuits, musical and anthropological. To my dad, who constantly encouraged me and kept my spirits up even when I felt like I would never finish. I owe my interest in sound and music entirely to my mom, who was a brilliant musician and lifelong piano teacher. Since I have been teaching full time for the past nine years, I have come to realize that my passion for and dedication to teaching is from watching her over the years. Her presence is imprinted throughout this book.

INTRODUCTION

Listening to the City: Distraction, Attention, and Ubiquitous Listening

CURRY SHOP MUZAK

On this particularly warm day in July 2012, while lingering in the popular curry shop chain CoCo Ichibanya in the Tokyo neighborhood of Takadanobaba, perhaps overstaying my visit in a place conducive to expediency, I found myself confronted by a gently cacophonous mix of sounds. The BGM (as background music, or muzak, is referred to in Japanese)[1] sounded like a Brazilian channel, for I heard the instrumental version of Antonio Carlos Jobim's 1962 classic "The Girl from Ipanema" multiple times during my visit. This repetition did not bother or surprise me, nor anyone else for that matter. No one appeared to be actually listening or responding to the music, which is not meant to be consciously heard. As is common with much background music, a jaunty violin backed by a soothing full string section replaced the familiar vocal line. The music was doing what it was designed to do: fill up space and enhance ambience without distracting customers. Layered on top of this music was the steady chatter of customers' loud voices (all men on this particular day), familiar Japanese phrases spoken by employees to customers, and the occasional burst of street noise and heat that punctuated the cool interior whenever the automatic doors neatly swept open. The incessant BGM loop became further entangled with the repetitive lo-fi jingles of a CoCo Ichibanya commercial played on a small boom box perched atop the ledge of the takeout window.

Tokyo is a richly dense and overwhelming sonic environment. Sound spills out of storefronts and demands the attention of passersby, who are confronted by it as they move through the city. These sounds create a complex and discordant public space that is sonically managed through constant supervision and curation by individual businesses. *Tokyo Listening* examines a range of distinct listening cultures that all reveal something about how listeners attune themselves to the sounds and noises of the city. The listening practices discussed in this book operate on a continuum that reveals an array of techniques for listening both *to* and *in* Tokyo. A classical music café regular whom I interviewed reflected on the process of listening in Tokyo, remarking that there are a variety of techniques, although he pointed out that one is not superior to the other. Listening privately at home without background noise in a controlled environment, listening intently in the music café, and reading in the café while the classical music operates as BGM are all potential auditory techniques that make up listening culture in Tokyo. The process of Tokyo listening is an attunement toward particular moods and affect, and is not dissimilar from attunement in other global cities. Tokyo listening invokes a mode of listening that is produced in and by the city: experimental music, music cafés, and muzak are all forms that are generated in urban spaces.

Tokyo Listening examines how atmospheres are sonically sensed and generated, an approach that entails a focus on the perspective of both the subject (the listener) and "the object, from the side of the agencies by means of which they are created" (Böhme 2017). Thus, my focus is on the techniques of affective management behind Tokyo's sonic textures, as well as on those who are subject to this music as they relax, listen, read, shop, and work. I track how the city atmosphere is acoustically staged for music listeners, shoppers, and office workers. Muzak is widely deployed to create calming atmospheres for shoppers, but it has also recently begun to be used to affectively manage white-collar workers in the hope of making them more productive. The transmission of sound evokes the smooth functioning of bodies, systems, and markets. The recorded sound of birds twittering in Tokyo train stations assists blind passengers who are moving through the station; office workers are surrounded by constant background music and ambient sound ("mental balance music for workers," according to one company's advertising to potential business subscribers); upscale department stores play ambient music in women's restrooms with the intention of creating a relaxing atmosphere.

At the opposite end of the sonic spectrum, experimental music listeners and classical music café patrons engage in an intentional mode of listening, seeking out sound that disrupts the managed sonic palette of the city. In contrast to those who hear publicly amplified sound deployed to evoke corporate normalcy, experimental music listeners engage with sound as a way to reattune themselves to the soundscape of the city. Similarly, rather than seeking silence, music café goers intentionally seek out a sound-filled environment in which to relax and drop out, a practice that contrasts with more common uses of cafés as efficient public spaces for work or creative labor, where (privately) listening to music is a by-product of the café, not central to it. Classical music cafés, primarily located in Japan's urban centers, are venues in which patrons drink coffee and listen to classical music recordings. While the modes of listening discussed in this book are all publicly shared modes of listening, they do not exclude private mobile listening, a mode of auditory engagement widely practiced in Tokyo and globally.[2] Thus, my book focuses on a type of listening wherein listeners are fully exposed to sound in public spaces such as department stores, cafés, offices, and music venues, rather than shutting out sound with private mobile technologies.[3]

Tokyo Listening explores how the urban sensorium actively forms the acoustic ground on which auditory practices are cultivated. The book maps out the linkages between listening practices and the city soundscape, examining how the sensory experience of the city informs how people listen to both music and everyday, ubiquitous sounds. Rather than push such noises to the side in order to examine musical sound, I argue that daily noises and ambient sounds serve as the foundation upon which listeners experience the city. In order to examine how people attune their ears to the sounds of the city, the book focuses on two distinct types of sonic practices: (1) spaces where music operates in the background and is not something people consciously engage with, and (2) spaces where patrons go to intentionally experience music. These two types of places in which sound is experienced cannot be understood apart from each other.

These auditory practices occur in such varied spaces as hotels, department stores, shopping arcades, white-collar offices, experimental music venues, and music cafés. I argue that listening practices and sonic engagement are refined, honed, and informed by the sensory experience of the city. Understood in this way, sound is integral to the affective management of listeners, consumers, and workers in their environments. In all of these cases, someone is deciding what is to be heard, whether that is the café owner, the sound designer, the office manager, or the musician. Significantly, the reception of sound operates along a continuum, suggesting the liquid nature of sound as it is managed and deployed in Tokyo. For example, in some spaces, sound is used to structure nonproductive leisure activities for distracted consumers or music listeners, while in other spaces, it serves to create a feeling of productivity and efficiency for post-Fordist office workers.

By definition, cities make sound. In the public spaces of urban Japan, characterized by Nakajima Yoshimichi as *oto zuke shakai* (sound-saturated society), sound is particularly prevalent. Everyday movements are routinely confronted and announced by sound such as subway melodies and announcements, white-noise machines in public restroom stalls, automated crosswalk jingles, raucous political announcements blaring through distorted loudspeakers, and ubiquitous background music, or BGM. The idea of saturation implies a sense of exhaustion, being overwhelmed and overcome by noise. An employee of a company that produces BGM with whom I spoke referred to Tokyo's soundscape as being "filled" with noise (*afurete*), a place overflowing with sound. A sound-saturated society is, to be sure, a contentious category depending on one's orientation to Tokyo's soundscape. Nakajima viewed such ubiquitous noise, and BGM in particular, as a form of "sickness." I queried a manager at USEN, one of the main corporate providers of BGM in Japan, about this idea, and he laughingly responded, "What's the problem with that type of society?" His response suggests that this saturation is not aberrant or offensive for someone who works to inundate the city with background music and sound. Indeed, it is the primary logic that sustains the corporate sensory manipulation of public space in Japan.

Tokyo's aural landscapes have been described by city dwellers as chaotic and cluttered, and some scholars have argued that residents have little control over being subjected to such sounds. City residents, it has been asserted, are unable to resist or recognize the dense sonic landscape that surrounds them (Yano 2005). Indeed, Japan is often said to possess a "cultural inclination" to accept noise in public spaces (Dolan 2008, 663). Rather than suggest that there is a cultural tolerance or predilection toward noise or ubiquitous sound, I situate my study of listening practices in relation to the affective management and corporate production of sound in contemporary Japan, a "national-cultural formation where as little as possible is left to chance" (Ivy 1996, "Mystery Man," 13). My premise is that soundscapes in Tokyo are not accidental or unintentional. The sensory attunement toward the city soundscape necessarily informs how people encounter and participate in collective forms of listening and living.

The lo-fi sonic atmosphere of contemporary Tokyo, much of which is created and produced by sound design companies, orients and informs the listener in everyday spaces of consumption. Sound design work in Japan hinges on a heightened awareness of public and private space and sensory attunement toward the acoustic conditions of such spaces. Tokyo's soundscape is primarily constituted by technologically mediated acousmatic sounds; that is, sounds we hear without seeing their source. This status helps explain their hold over the city's inhabitants. While many of these sounds are in the form of disembodied voices admonishing warnings to pedestrians or train riders, one of the more audible features of Tokyo's soundscape is BGM, perhaps the quintessential acousmatic musical form. Background music and ambient sounds are produced and circulated throughout Japan's urban spaces to a heightened degree.

City noises have historically served as sonic markers of progress and modernity, from street barkers to the sounds of industrialization. However, by the early twentieth century, this perspective shifted to viewing noise as a nuisance and public health problem similar to smoke pollution. The American noise abatement movement was first formed in 1906, and it helped frame city noise as antithetical to progress and civilization, resulting in new legal measures to regulate and control noise. Emily Thompson describes how such abatement measures slowed down in the wake of the Great Depression but reemerged by the 1970s, during a period in which noise pollution became infused by a sense of "environmental consciousness" and was tackled by grassroots organizations and ordinary citizens (2005, 197). She notes that in the postindustrial city, where consumption is more predominant than production, noise abatement tactics are largely determined by issues related to "quality of life" rather than the environment or matters of efficiency (2005, 198). Today, consumers are able to shut out noise with private mobile technologies such as headphones. Although, as we see from a contemporary poster for a Tokyo Metro anti-noise advertising campaign, even headphones do not provide enough space from "sound leakers."

A recent series of subway etiquette posters for Tokyo Metro features the expected admonitions, including those asking passengers not to take up more than one seat and to be careful when bringing a wet umbrella into the train. One such poster, shown here, stood out to me on a recent trip to Tokyo. The top of the poster reads in English and Japanese, "Are your manners in good shape?" Underneath that heading is a cartoon drawing of a young man, clad in ripped jeans and what appears to be a leather jacket, wearing oversized headphones that he is pressing against his ears with both hands. The passengers seated next to him are annoyed; one is covering his ears, and another has her arms folded in frustration. Underneath the drawing it reads, "Sound: Please take care that the sound does not leak from your headphones inside the train." The poster's basic

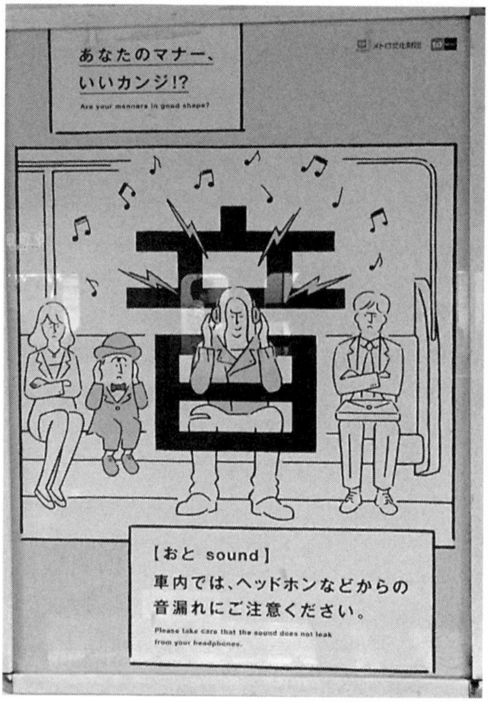

FIGURE I.1 “Are your manners in good shape?” Subway etiquette poster on the Tokyo Metro. Photo by the author.

theme is sound, made visible through the Chinese character for sound (*oto*), which occupies the entire frame of the headphone leaker. In fact, the young man is drawn as if he is actually emerging from the character itself. There are yellow lightning bolts and a cluster of music notes dancing above the character for sound.

By highlighting the social dangers of being a “sound leaker,” this poster frames sound as liquid material to be contained and restrained in the public space of the train.[4] It also acknowledges the territorializing force of sound. Gilles Deleuze and Felix Guattari’s notion of the refrain (*ritournelle*) evokes this potential whereby sound fills up space and demarcates a territory or order: “A housewife sings to herself, or listens to the radio, as she marshals the antichaos forces of her work. Radios and television sets are like sound walls around every household and mark territories (the neighbor complains when it gets too loud)” (1987, 311).[5] Given the promiscuous presence of noise throughout Tokyo—muzak emanating from stores, voices on trains and near escalators admonishing pedestrians, store barkers screeching through megaphones at the store entrances, among other everyday sounds—this subway poster seems to suggest that there are certain spaces (however few) in Tokyo where ubiquitous noise is not sanctioned. Listening in these spaces should be private and controlled. Nevertheless, this subway etiquette poster reveals the ways in which sonic practices more generally are undergirded by rules and logic: namely, control your sonic leakage.

Tokyo Listening focuses on intimate public spaces of listening where people are fully exposed to amplified sound, whether recordings or live performances. I focus on three sites of attunement to sound: experimental music venues, music cafés, and retail and work spaces targeted by BGM. By studying these places of sensory attunement, we learn how people in Tokyo confront the power of the broader city soundscape through their everyday sonic engagements. While we might expect people to seek out quiet spaces or use private audio technologies in order to shut out city noise, *Tokyo Listening* reveals that people seek out sound-filled spaces in order to mitigate against such city noise by focusing on the sensory experience of urban listening as a planned and multifaceted dimension of everyday city life. Rather than disengage from the sounds of the city, these auditory practices reflect and respond to the everyday liquid rhythms of urban life in contemporary Japan. Such acts of listening complicate the division between processes of attention and distraction as critical to urban sensory engagement. As my analysis of listening practices in experimental music venues and music cafés reveals, residents of Tokyo hone their listening skills through the sonic distractions of the city.

My focus on public forms of listening in relation to the sensory space of the city expands upon recent work on similar topics, which has tended to focus on listening and mobility as solitary urban experiences. For example, Michael Bull's *Sound Moves: Ipod Culture and Urban Experience* (2007) frames listening within everyday urban life as a necessarily private (and aestheticizing) experience wherein listeners seek to disengage and shut out the noise of urban life through mobile technologies such as the iPod or mobile phone. *Tokyo Listening* broadens the scope of such recent scholarship on privatized listening to focus on aural practices that are necessarily informed by the sensory experiences of the city, where listeners are fully exposed to sound.

In this book I understand the process of listening not merely as an auditory process, but as an embodied practice and sensory attunement. Despite the fact that it remains commonsensical that humans hear with their ears, many scholars have argued that people do not take in sound simply through our ears, but through our bodies as well. Listening and hearing, in short, are intensely corporeal processes. The listening practices examined in this book all contribute to this understanding of listening as not merely an auditory and single-sense practice. As Don Ihde puts it, "the ears may be focal 'organs' of hearing, but one listens with his whole body" (2007, 135–36). It is critical to recognize that the senses are "multiply related" and necessarily act in concert with one another (Connor 2004, 153). Thus, humans never interact with or confront our surroundings via only one sense. Sensory perception, as scholars such as David Howes have argued, must be examined through intersensoriality, or the "multi-directional interaction of the senses and of sensory ideologies" (2005, 9). Our auditory perception, as Steven Connor points out, is also inherently tactile. He describes the paradox of hearing as "at once intensely corporeal—sound literally moves, shakes, and touches us—and mysteriously immaterial" (2004, 154). The tactility of sound, Connor explains, is deeply dependent on sound's disembodied quality.

Tokyo's dense sonic environment gives rise to a range of perceptual responses. Brian Kane's analysis of acousmatic listening to the city offers a provocative lens through which to grapple with these responses. He describes it as a "shared, intersubjective practice of attending to musical and nonmusical sounds, a way of listening to the soundscape that is cultivated when the source of sounds is beyond the horizon of visibility, uncertain, underdetermined, bracketed, or willfully and imaginatively suspended" (2014, 7). Sound design companies, for example, seek to preserve the acousmatic nature of such sounds in order to tap into the listener's senses and moods. One Tokyo-based sound design company that I interviewed has recently begun to incorporate scent into its marketing strategies and sound design programming.

In addition to BGM listening (or non-listening), I examine related modes of listening and sensory attunements toward sound and music in Tokyo: experimental music listening and focused listening in classical music cafés. Along with BMG listening, these auditory practices are affectively shared listening experiences. Such modes of listening are meaningful precisely because they are cultivated in relation to the chaotic and obtrusive sonic environment of the city (Kelman 2010, 230). Moreover, these listening practices are all deeply intersensorial and embody processes of engaging with and responding to sound. For example, in Chapter 3, I examine how white-collar office workers in Tokyo who are subject to constant music programming at work engage with BGM in a very particular (and highly corporeal) manner that dissuades them from actively listening to it. This use of BGM exploits somatic modes of engagement with sound other than listening, including the workers' bodily movements, neuroactivity and biorhythms, and sonic cocooning. Unlike workers in North America who create personal playlists as a privatized form of mood enhancement, office workers in Japan are encouraged to refine and nurture their emotional attachment to work and their co-workers through this shared and embodied listening experience. Workers are presumably more efficient and, by extension, happier. Themes of productivity (and nonproductivity), happiness, and comfort run through all the chapters of this book, such that sound in Tokyo produces affective states designed to ease participation in modern life.

TOKYO AS "ADVERTISING CITY"

As Japan's major urban center, Tokyo is variously imagined as chaotic, refined, and futuristic, even verging on the hyperreal. Joseph Hankins and Carolyn Stevens point out how Tokyo's density complicates stereotypes about a supposedly Japanese affinity with nature as a "cultural virtue" (2013, 9).[6] As a result of Japan's rapid postwar economic growth, beginning in the 1970s, Tokyo's urban spaces became reimagined as what Yoshimi Shunya referred to as a type of "theatre." In this case, Yoshimi was referring to the marketing tactics of the two leading department store conglomerates, Seibu and Parco. Urban space was staged in order to appeal to a segmented consumer market: "individual urban spaces were not just put together as a series of products acting as signifying code for a particular taste; they had to be created as a stage where visitors could act out their roles according to these codes." Department stores were thus designed as a "linked series of compartmentalized, multi-layered, themed spaces" (Yoshimi 2011, 289), similar to Disneyland. As Yoshimi argues, the visitors to this advertising city were also readers, and were "newly positioned as a viewing apparatus to *see* and at the same time *be seen*" (2011, 290).[7]

Akiko Takeyama explores how the Japanese government was invested in marketing Tokyo as a futuristic city in the 1980s during the bubble economy in order to align Japan with the global economy and move past the domestic social and environmental ills of the 1970s (2016, 26). Takeyama notes that Kitada Akihiro described Tokyo as an "advertising city" where "everything becomes a potential advertisement" (2016, 29) and the consumers themselves serve as publicity agents for luxury brands as they move through city spaces. Less discussed by Kitada and other scholars, however, is the role of sound in these spaces of hyperconsumption, as well as the possibility of noncorporate-sanctioned sounds. Of course, sound penetrated these spaces, but it was not "seen" by theorists. The omission of sound in theories of city space underscores how the ubiquity of sound could simultaneously make it overlooked. The dynamic uses of sound design techniques that I examine in this book are conducted in the service of consumer capitalism: sound is used to generate affects and produce sociality.

Indeed, the tactical role of sound in contemporary practices of hyper-consumption prompts us to reevaluate how the sonic environment was managed in the recent past. Historian Jordan Sand examines the historic preservation movements that emerged in the 1980s during Japan's bubble economy era, describing them as an impulse to preserve the "vernacular" spaces and fragments of Tokyo's past that somehow escaped the rapid shift toward redevelopment and hyper-urban planning beginning in the 1970s. The desire to maintain these vernacular spaces—alleyways or low-rise wooden structures that were seen by developers as inefficient uses of land—begs the question: Is there a sonic vernacular? If we can conceive of buildings as palimpsests, is there an auditory analogue? Are there sounds that linger despite having no apparent "efficient" use-value?[8] I explore this question in my chapter on music cafés and their associated modes of listening, their design, and the sounds they propagate. Nostalgic sounds are also a constant, if curated, dimension of BGM programming throughout the city. As amplified sound in Tokyo becomes increasingly corporatized—BGM in consumer spaces, train station melodies that mimic local advertising jingles such as the Ebisu beer brewery—the possibility of vernacular sound becomes an increasingly timely question.[9] I suggest that such managed sound in Tokyo serves as a kind of auditory hyperesthesia, drawing on David Howes's argument that capitalism seduces "the senses of the consumer in the interest of valorizing capital" (2005, 289–90).[10]

Sound has long been used in commercial transactions in Japan. This is especially true in the case of store vendors calling out to customers, an act known as *yobikomi*, which serves as a way to sonically beckon potential customers passing by. While these vernacular sounds, which have shifted from being chanted or performed live to amplified recordings featuring musical jingles, might be seen as in danger of disappearance, Carolyn Stevens points out that the street continues to serve as a vital site for sonic retail practices (2016, 98), yet also that in today's Japan, there are varying orientations toward the sound of the vendors. For example, the amplified call of mobile vendors selling roasted potatoes from the backs of their trucks is pleasurably nostalgic for many, similar to the "tinkling jingle of the ice cream truck drifting through the air on a summer day" (2016, 95). The opposite is true of the sound of vendors selling bamboo poles for drying laundry on one's veranda. Namely, these vendors have come under scrutiny for fraudulent advertising and overcharging customers (2016, 96), and as a result, are seen in a more negative light. Consequently, many residents of Tokyo complain about their "excessive noise."

INTENTIONAL LISTENING

In this book I seek to bring together two different types of listening spaces in the same frame of reference: spaces people *go to* specifically for the music, and places where the music *comes to* them. The structure of this book is designed to situate these two types of listening experiences alongside each other, since they both constitute significant aspects of the city's soundscape. I begin with semiprivate esoteric spaces (experimental music venue, classical music cafés) where people go to intentionally listen to music. Then I move outward to public spaces of consumption (department stores, shopping arcades, public restrooms) and labor (white-collar offices in Tokyo); spaces where the people are not consciously listening to the background music. I devote Chapter 1 to exploring the relationship between experimental listening practices and sensory attunement toward city sounds. In this chapter, I examine the experiences of listening to music at Off Site, a small experimental music venue in Tokyo whose unique acoustic environment led the audience to absorb the music along with sounds of the city outside.

Chapter 2 addresses the relationship between intentional listening and the experience of the city in a very different space: classical music cafés. While both jazz and classical music cafés are known for fostering serious listening, as occurred at Off Site, I argue that contemporary classical music cafés provoke a peculiar form of listening to and engagement with sound. Unlike workers who use subtle background music as a means to improve productivity and provide calm (Chapter 3), the impulse in the music café is to move toward a sound-filled environment in order to relax. Off Site listeners, on the other hand, drew on the sonic distractions of the city in order to become more concentrated music listeners. Some music café patrons will listen in a distracted manner and use classical music as BGM while they read, sleep, eat, or stare into space. Deeply nostalgic spaces for older patrons, classical cafés are now also key sites for younger patrons to listen to recorded music (always vinyl) at a high volume in public space, thus a rejection of the private auditory experience that is often characterized as the primary mode of listening in contemporary urban public space.

Significantly, classical music cafés are also comparatively quiet and contemplative public spaces for people to engage in nonlaborious solitary activities while enveloped by sound. Music cafés foster an environment with strict rules of etiquette, including no talking and often no cell phones or laptops. The leisurely ambience of these cafés stands in stark opposition to the hyperefficiency of chain coffee shops or convenience stores. The process of listening at classical music cafés is a highly embodied and intimate experience that is cultivated by the café's antiquarian atmosphere and design. Classical music cafés are, I argue, deeply bound to an atmosphere grounded in patina, an expression of the "aestheticizing effect of time" (Diaconu 2006, 131). Often explained as "the visible surface of a temporal depth" (2006, 132), patina is also a highly tactile and sedimented process because it "makes visible not only the subject's corporeality, but it embodies also a certain 'vulnerability' of the material and its memory" (Diaconu 2011, 29). Chapter 2 explores the use of analog sound as sonic patina through which café owners are able to create a nostalgic and slow-paced atmosphere: an unproductive, yet comfortable place that holds the city at bay.

Moving from semipublic and rarified spaces to public spaces, Chapters 3 and 4 explore the recent use of BGM in two very different contexts: the white-collar workplace and retail environments such as department stores, restaurants, and hotels. Chapter 3 examines how BGM has recently been deployed to manage and care for over-stressed office workers at a time in which challenges and questions about mental health issues, work-life balance, and the negative effects of overwork are on the rise in Japan and elsewhere. Sound, in this context, is used to relax workers while simultaneously making them more efficient and strengthening their attachment to the workplace and co-workers. I tie this use of BGM in the workplace to the broader history of muzak as it was used in North America during the post–World War II era in order to stimulate workers and keep them happy and productive.

In Chapter 4, I situate BGM in Japan in relation to the global history of background music and mood management, exploring in particular historical and contemporary convergences and points of disconnection. I explore the "muzak'ed" saturation of retail and leisure spaces such as convenience stores, bars, restaurants, shopping arcades, and department stores, and examine how background sound is deployed to create affective atmospheres and sonic forms of branding. I position BGM as a form of sonic air conditioning to evoke the infrastructural nature of background music as a kind of ambient sensory engineering that people are not supposed to notice or listen to directly. In this chapter, I complicate assertions of muzak listening as a necessarily passive activity and reveal how claims by sound design companies that people are not supposed to listen to BGM are belied by their extensive design techniques specifically meant to target patrons on an auditory and sensory level. The Japanese public is subject to this constant sonic intrusion of BGM as they commute, shop, engage in leisure time, and more recently, as they work. As I discuss in Chapter 4, although some commuters might appear to tune out BGM through headphones, there are others who actively listen to Tokyo's background music.[11]

Muzak is a musical form that is born of the city—it masks, mimics, and fluidly smoothes over the noise of the urban soundscape. Much of the discourse on muzak is bifurcated: background music is viewed as either an aesthetically insipid form of music or a sonic tool of corporate manipulation. In this book I take the production and consumption of BGM in Japan seriously in order to avoid condemning muzak on aesthetic grounds.[12] Some research interlocutors in Tokyo whom I interviewed surmised that BGM has become a necessary form of customer service for businesses in Japan's major cities, thus implicitly suggesting that sound is deeply entangled with productivity itself. They also suggested that BGM is much more prevalent in cities, and not nearly as ubiquitous in rural areas. The logic that undergirds BGM programming and its accompanying discourses of listening engagement is audible throughout a wide variety of Tokyo's public spaces, ranging from conventional spaces for background music, such as department stores, supermarkets, and restaurants, to private spaces in public, such as restrooms. In Chapters 3 and 4, I examine BGM as meaning more than simply "background music," but background *sound* more generally. Thus, I include a wide variety of sonic devices under the rubric of BGM. These include white-noise machines in women's public restrooms, piped-in sounds of the Borneo rainforest near department store escalators, train melodies, and old-school muzak piped through retro shopping arcades. All of these BGM devices operate as planned and programmed elements of the sensory experience of the city and its built environment.[13] The widespread deployment of BGM throughout Tokyo is predicated on the idea, held as truth by some sound design companies, that spaces without sound are experienced as inherently unsettling and "lonely" (*sabishii*). According to company lingo, sound serves to keep businesses (and patrons) "advancing."

Like smell, sound is elusive (Tsing 2015, 46). It is notoriously difficult to define and describe. It is even harder to ask other individuals to explain how they engage with sound. Asking service workers for their opinions on BGM is a little like asking someone what she thinks of the air conditioning or heating system. It is a query fraught with challenges, and yet it remains a productive and necessary ethnographic impulse. Sound (like smell) is "a sign of the presence of another" (2015, 46). Ethnographic methodologies reveal the contradictions and gaps between what BGM and sound design companies say music does to spaces, and the experience of the people who inhabit spaces and are subject to the music's infrastructural presence. In the chapters that follow, I linger on the social relations (specifically, public forms of intimate listening) that cohere in the spheres in which sound is used to affectively manage or comfort the city's inhabitants.

LISTENING TO THE AVANT-GARDE

Research for this book is partially based on participant-observation fieldwork with Tokyo's avant-garde music scene, beginning in 2000. I conducted three continuous years of dissertation fieldwork from 2003 to 2006, and I have made multiple return trips since then to update and expand upon my original research. Soon after this research began, I became more attuned to the broader city soundscape in which these performances were occurring.[14] This connection between experimental music and city sounds is not surprising, given the long history of the avant-garde's interest in the sensorial and synaesthetic experience of the city. I conducted ethnographic research in three broad areas of Tokyo: experimental music and art scenes, classical music cafés, and BGM and sound design companies. I regularly attended performances at experimental music venues throughout Tokyo, including minimalist electronic improvised music, noise music, punk, jazz, and classical. Early in my fieldwork, I studied at Bigakkō, a long-standing art school in Tokyo, under artist and musician Itō Atsuhiro. Through these studies, I was able to build important networks and connections with the experimental music and art worlds of Tokyo. I interviewed experimental music listeners, musicians, live-space owners, music critics and scholars, and the owners and employees of record stores. Through my fieldwork, I learned that sound in Tokyo was being employed to manage, heal, and sustain city dwellers as they worked, shopped, and lived their lives.

ONE

Learning to Listen to *Onkyō*

Ear Training as Sensory Attunement

The first time I entered the now-defunct live performance space Off Site was the first time I had ever felt on edge as an audience member at a music venue.[1] This was partly due to the extreme proximity of the audience to musicians in this tiny space. When I sat in the front row, I felt that if I leaned forward, I could touch the empty turntable as it spun and projected indeterminate and abstract sounds. But this tension was also due to the sonically extreme nature of the music that occurred in the space. Founded as an alternative to Tokyo's live house music system, where musicians must pay to rent the space and often go into debt in the process, Off Site promoted experimentation in both performance and listening. The music at Off Site came to be referred to as *onkyō* (literally "sound," in Japanese), a minimal, improvised genre of electronic music that emphasizes sound texture, gaps, and silences rather than melody or rhythm. Rejecting the pay-to-play model, Off Site's experimental philosophy was engendered, on some level, by the unexpected material and acoustic dimensions of the space itself. With its blank white walls and concrete floors, Off Site invoked the white-cube gallery model. Because of the close proximity of the building's neighbors and the poor sound insulation, all live performances had to be extremely quiet. So quiet that listeners and musicians soon began to notice how outside sounds (what John Cage called "non-intentional" sounds) leaked into the space and would often commingle with—and sometimes overpower—the "music itself." Fragments of conversations of passersby; a squirrel scrambling and darting about just outside the front entrance; noise from the nearby train station; ambulance sirens in the distance. All these sounds were heard inside the venue.

The corporeal, and sometimes strenuous, engagement with sound was a prominent theme that emerged in discussions about listening at Off Site. A music reviewer for the British avant music magazine *The Wire* once described the intense listening at Off Site as "listening that you can almost see, suspended in the room like colored light." Japanese music critics described the act of listening at Off Site as a process of "straining one's ears" (*mimi wo sumasu*). Listeners, myself included, seemed to relish being a part of this community of "difficult music." The unexpected non-intentional sounds inside the performance space conjured up a sonic experience that was antithetical to the corporate-sanctioned sound that surrounds Tokyo's residents on an everyday basis. This was not ordinary listening. And yet, somehow, ordinary sounds (traffic, fragments of conversation, squirrels scurrying) entered into the space and provoked sensory realignments. It was this intense auditory process, operating as a kind of sensory training, that caused some listeners to discover that their senses had been newly attuned toward the urban soundscape. The attunement in this case operated as "an affective orientation or register" (Peterson 2016, 96). Affect has been defined by Brian Massumi as "intensity" and "sensation," something that precedes emotions, which, as Michael Hardt and Antonio Negri point out, are "mental phenomena," while "affects refer equally to body and mind" (2000, 293).

Many listeners told me that they heard the city differently after attending shows at Off Site. Intriguingly and perhaps paradoxically, they became distracted city listeners through the contemplative space of the white-cube gallery model. Here, focused, serious listening unwittingly converged with the sonic distractions of the tactile urban landscape. Because of the performance space's restrained and restrictive atmosphere and the difficult nature of the music itself, listeners often spoke of having to "persevere" in order to maintain bodily focus and attention. Many of them developed listening techniques and strategies to cope with the bodily effects of straining one's ears during live performances.

ENTERING OFF SITE

Immediately upon entering Off Site, audience members encounter the owner, Itō Atsuhiro, seated at the back of the stark, ground-level performance space, selling tickets. As is customary in Tokyo, one drink, often beer, wine, soda, or oolong tea, must be purchased along with the ticket. Drinks are served on the second floor by Itō Yukari, Itō's wife at the time and Off Site's co-owner. Tickets are typically 1,000–1,500 yen, considerably cheaper than tickets for most live music performances in Tokyo, which can run upwards of 3,000 yen. Inside the performance space, the audience members sit on narrow aluminum-backed folding chairs that force them to perch uncomfortably close to their neighbors, evoking a form of bodily tension akin to a subway commute during rush hour. This tension is inherent in *onkyō*'s tense performance environment at Off Site, in which listeners must strain their ears (*mimi wo sumasu*) in order to grasp traces of *onkyō*'s barely audible presence.

Those already familiar with Off Site's unique ambience and cramped conditions arrive early and claim seats in the first row. At most experimental performance spaces in Tokyo, performers and listeners are seated throughout the performance, thus imposing a sense of order on the spatial dimensions of the live music space. This arrangement imparts to the listeners an air of detachment. Seats are arranged neatly in four to five per row, and are separated by a narrow, barely passable aisle. Maximum capacity is between thirty and thirty-five, though it is not uncommon for a performance at Off Site to draw fewer than ten listeners. Because there is no stage and the seated musicians perform on the same level as the audience, sitting in the back means limited visibility. Listeners wait in somewhat eerie silence, mostly limiting conversations to soft, hushed tones. Most sit rigidly in their chairs, quietly waiting for the performance to begin.

After lounging on the second floor, the musician(s) enter the performance space. A subtle hush falls over the room. Aware that the musician(s) will generate a small amount of noise by shifting or setting up equipment just prior to the performance, many listeners take advantage of this rustling, however minute, to clear their throats or take a final gulp of a drink before gently placing it on the floor underneath the seat in front of them. They do this because once the performance begins, the listeners must follow unwritten codes of etiquette, including maintaining silence and concentrating closely on the music while avoiding any outward reactions. The tension is heightened at Off Site because, unlike in most music or theater venues, the boundaries that mark the beginning of a performance at Off Site are not readily apparent—the listeners cannot simply sit back and relax. There is a thick atmosphere of uncertainty and unrest as the audience's silence is intensified by its impending confrontation with a performance that is marked by minutiae, sparsity, and long spatial gaps. Its silence will also be confronted by quotidian city sounds filtering into the performance space.

This chapter investigates the linkages between experimental listening practices

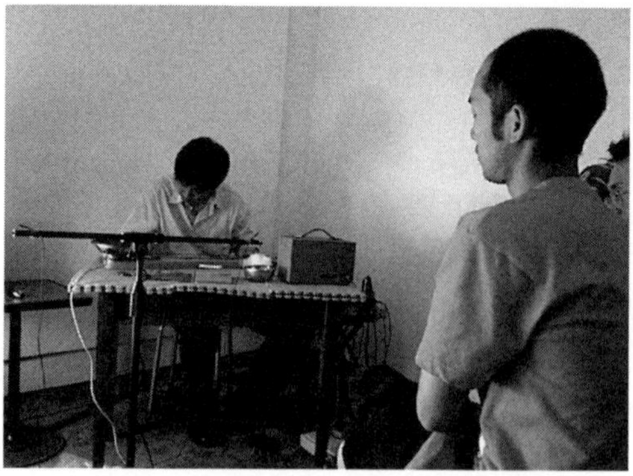

FIGURE 1.1 Off Site audience member with eyes closed. Photo by the author.

and the sounds and noises of urban space in contemporary Tokyo. I situate *onkyō* listening practices in relation to discourses of putatively "Japanese" modes of listening, as well as culturalist explanations for the general acceptance of noise in Japanese public spaces. How are *onkyō* listening practices—often framed as "serious" (*majime na*) listening—informed by the distracted environment of the city and the listening habits that the urban soundscape engenders? Although previous researchers have characterized the Japanese public as unable to resist or "assume control" over urban noise (Yano 2005, 196), many of the musicians and listeners I spoke with revealed a critical engagement with Tokyo's soundscape.[2] Rather than passively accepting Tokyo's dense soundscape without reflection, listeners and musicians in Tokyo's experimental music scene were highly conscious of the impact of the urban soundscape on modes of performance and listening, as well as on everyday life more generally.

How might the urban soundscape facilitate or otherwise influence the development of new auditory practices? Sensory engagement with the city has historically taken on numerous forms, including through the quintessential (and always marked as male) figure of high modernism, the *flâneur*, who aestheticized public space through his detached gaze. Michael Taussig discusses how Walter Benjamin likened the bombardment of advertisements and billboards confronting the distracted urban walker to the tactile effect of Dadaist cinema (Taussig 1991). If subjectivity and selfhood in modernity are defined by the distracting and fragmenting shock effects of navigating the urban metropolis, then how are we to explain the current sensorial engagement with the contemporary city? How do people make sense of navigating everyday life—including aurality—in a "sound-saturated" environment such as post-economic-bubble Tokyo?

I examine the relationship between Tokyo's dense and intrusive soundscape with the emergence of modes of listening that produce refined or heightened senses, enabling the listener to hear everyday sounds in a more critical fashion. My analysis draws on an examination of the demands that *onkyō*, a form, places on the sense of hearing and the act of listening. I explore listeners' assertions that their sense of hearing has changed as a result of listening to *onkyō*, including the broader implications of losing and regaining one's senses and the disciplinization of aurality in recessionary Tokyo. This refers to the ubiquitous presence of BGM in Japan, a sound that is not meant to be listened to actively, but to be heard only as background. What sort of social and spatial conditions must exist for listeners and musicians to notice a change in their hearing, and, implicitly, a heightening or refining of the senses?

By considering the presence of everyday sounds in Tokyo's soundscape, this chapter engages with the highly subjective nature of sound, particularly the politics of the urban soundscape. When does sound become noise? Noise is generally acknowledged to be that which disturbs, disrupts, and interferes; it is a sound or message that we do not want to hear. It is excessive and uncontrolled. Information theorists have considered noise to be an interference or disturbance in transmission—"a signal the sender does not want to transmit" (Moles 1968, 78–79). Acoustical science, however, interprets noise as that which is opposed to sound, the antithesis of "desirable" sound. In addition, urban noise is often equated with crowds and the masses; thus pointing to noise's potential as a social threat.[3] Ultimately, however, the border between sound and noise (or music and noise) is a highly subjective demarcation, and often culturally contingent.[4] The ambiguous divide between "musical" sounds and city sounds during *onkyō* live performances creates an environment leading to what some listeners have referred to as "changed hearing" (*mimi ga kawatta*). This sensory attunement illuminates the complex relationship between distraction and contemplation in urban settings, including the possibility of refining sensory perception and awareness. By focusing on the co-constitutive dimension of live performance sounds with that of the aural landscape, this chapter explores new modes of perception in relation to Tokyo's soundscape and the listeners' engagement with sounds of the urban everyday.

MODERNITY AND SENSING THE CITY: DISTRACTION, ATTENTION, AND SHOCK

Central to understanding listening practices and aurality in the contemporary city is an understanding of how the senses have been hierarchically ordered in modernity. Modernity is frequently characterized by a regime of vision in which the gaze comes increasingly under surveillance. The question of control, surveillance, and the disciplinary gaze also speaks to the increasing disciplinization of aurality in modernity. An obvious example here is muzak, a form of corporate-sponsored programmed music that Jacques Attali views as "contributing to the establishment of a system of eavesdropping and social surveillance" (1985, 8). For Attali, such mass-mediated music, with its explicit linkages to practices of mass consumption, serves to silence everyday noise through repetition.[5] It is perhaps more accurate to think of the sensory effects of modernity as a "vast reorganization of sensory experience" (Hirschkind 2004, 131), facilitated by the development of new technologies such as film, the railway, and the expanding role of advertising and consumer culture.[6] The ocularcentrism of modernity is characterized as one in which social actors are increasingly subjected to surveillance and the disciplinary gaze; I would argue, however, that sound and aurality are similarly imbricated in modernity, in conjunction with the eye. Drawing on work by David Howes, I examine sensory perception via intersensoriality. As I will show, however, this is not to suggest a stable egalitarian relationship between the senses: the senses are not bounded discrete entities.

Modes of aurality and visuality are both implicated in the sensations and bodily experience of modernity. The visual regime, however, has typically been characterized as a Western, and thus implicitly superior, faculty; aurality has generally been interpreted as the lesser and often "non-Western" faculty (Hirschkind 2006). To put it plainly, the ear has long been viewed with suspicion and derision as the lesser of the senses. Despite this sensory hierarchy, and the fact that listening and hearing reflect distinct processes, their differences are nevertheless often elided and explained interchangeably as the same process or faculty. Jonathan Sterne discusses the distinctions between listening and hearing, noting that "listening is a directed, learned activity: it is a definite cultural practice. Listening requires hearing but is not simply reducible to hearing" (2003, 19). Sterne explains that the difference between the two faculties lies between activity and capacity. The intersection of *onkyō* listening practices and the intrusion of the cityscape suggests that the process of listening hinges on a learned ability to listen properly. Listening in this context requires attentiveness not only toward (musical) sound, but also toward the distractions and intrusions of urban sound more broadly. The mode of listening to *onkyō*—in which the listeners physically and sonically detach themselves from the continuous aural stimulus of the city—demands attention and concentrated listening on the part of the spectator, yet the audience members' contemplative listening is confronted with the sounds of the city spilling through the walls of the performance space. Such acts of listening perhaps complicate the division between processes of attention and distraction as critical to urban sensory engagement.

The experience of modernity has long been framed as one of distraction, fragmentation, and shock, though scholars and theorists have disagreed over the nature of such forces, whether they are ultimately liberating or "regressive" for the individual.[7] For Walter Benjamin, distracted perception was a liberatory outcome of mechanical reproduction. Moreover, it surpassed earlier modes of aesthetic reception, primarily reactions to paintings, which would "invite the speaker to contemplation" (1968, 223). Whereas prior to technological reproduction, a painting would "invite the spectator to contemplation," a film literally overcomes the spectator and forces him to confront it, thereby radically refiguring and "interrupting" the spectator's "process of association" (1968, 238). In contrast to these earlier contemplative modes of perception, the film spectator has no choice but to respond in a distracted manner, due to the disconnectedness and velocity at which the images are projected. Contemplation and distraction regarding visual perception, I argue, can be applied to aural perception, especially in relationship to the everyday experience of the city.

Michael Taussig discusses Benjamin's idea of tactility and distraction for the urban pedestrian, noting that he took a "cue here from Dadaism and architecture, for Dadaism not only stressed the uselessness of its work for contemplation, but that its work 'became an instrument of ballistics'" (Taussig 1991, 149). With its radical and violent techniques of dissemination, Dadaist art "hit the spectator like a bullet, it happened to him, thus acquiring a tactile quality" (Benjamin 1968, 238). For Benjamin, the sense of distracted reception was inextricably linked with the experience of walking in the modern city, "the type of flitting and barely conscious peripheral-vision perception unleashed with great vigor by modern life at the crossroads of the city, the capitalist market, and modern technology" (Taussig 1991, 148). Benjamin does not directly engage with the aural dimension of the modern city that confronts the distracted walker, but rather points to a sense of tactility emerging from distracted vision, in this case, the urban *flâneur*'s engagement with architecture, film, and advertisements (1991, 149).[8]

Can we speak of a tactility of sound that the contemporary pedestrian encounters, akin to Benjamin's discussion of the tactility of distracted vision as a constitutive sensation of urban modernity? Given how many pedestrians and commuters drown out city noise with headphones—yet who also often aurally disturb those in close proximity, like the "headphone leakers" on the Tokyo Metro—are such practices indicative of "ubiquitous listening" (Kassabian 2001, 2013) an effect of the growing omnipresence of music and sound in everyday life? How is this form of distracted listening different from the distracted vision that Benjamin spoke of in relation to capitalist modernity and advertisements in the city streets?

ONKYŌ AS EXPERIMENTAL FORM AND OFF SITE AS EXPERIMENTAL MUSIC SPACE

As my ethnographic fieldwork revealed, ubiquitous and distracted listening plays out in the context of experimental music in sound-saturated contemporary Tokyo, a city both championed and bemoaned for its intrusive soundscape. It is difficult to precisely trace the emergence and contours of the genre known as *onkyō*—improvised music that not only is sonically ambiguous and difficult to categorize, but also features practitioners who themselves vehemently reject their supposed membership in the genre.[9] It is important to note that the term *onkyō* did not emerge out of a consensus by the musicians and performers involved. Rather, it is a Japanese word referring to sound and the sonic that was first used by outsiders to describe the genre. Yet as many Japanese music critics and musicians have argued, to refer to a musical genre—which is always already defined by sound—by the word "sound" is both tautological and repetitive. Musicologist Hosokawa Shuhei likens the emergence of *onkyō* as a distinct genre in the 1990s to the development of the music category of J-pop that first emerged at HMV records (n.d.). Indeed, it is often said that the term *onkyō* first emerged as a genre category at Paris-Peking Records in Tokyo introduced by the store's owner. Musician Otomo Yoshihide was later credited with using the term (which he now laments) while on tour in Europe, at the time referring to improvised music sessions in Tokyo.

By 2000, with the opening of Off Site, a free gallery, performance space, and bookstore/music shop in Tokyo, *onkyō*, as it was now loosely termed, would soon find an aesthetic home. During Off Site's five-year existence, the notion of *onkyō* and its accompanying discourses of "attentive" or "serious" listening gradually began to be circulated and consumed throughout Tokyo's experimental music circuit. Many public talks and discussions in this period focused on themes seemingly spurred by Off Site's experimental aesthetic, such as notions of listening (*chōshu*), sound, and space that were believed to be fostered by the performance venue. In the process, the very notion of "listening at Off Site" began to be imbued with meaning and consumed, much to many performers' chagrin. As foreign media coverage of *onkyō* increased, listeners in Tokyo began to take note of Off Site's presence. By Off Site's final year, in 2004, as curious listeners began to increase in number, many performers and audience members had come to bemoan a lack of spontaneity and experimentalism that had characterized Off Site before it became known within the international experimental music community.[10]

As a minimal form of electronic music that demands a heightened sense of listening, *onkyō* emphasizes sound texture, gaps, and silences rather than melody or rhythm. Instruments used ranged from the traditional—saxophone, guitar, and harp—to the more experimental. Some performances included turntables without records, an empty sampler, and even an amplified fluorescent light tube that emitted soft flickers of sound in conjunction with the unstable flashing of light penetrating the audience's eyes. The majority of music performed at Off Site was referred to in Japan as improvised music (*sokkyō ongaku*), and yet, the form known as *onkyō* became inextricably linked with the space. Citing Veit Erlmann, Marina Peterson writes that experimental improvised music suggests the "global imagination" that arises from a "cosmopolitan, democratic and free sociality" (2007). Despite this, *onkyō*, with its attention to silences, pauses, and gaps, was continually evoked as evidence of a quintessentially "Japanese" aesthetic, regardless of the strong transnational presence of improvisational musicians, many of whom collaborated with Off Site's regular performers.[11] As I argue, *onkyō* performance and its associated listening practices, which led many listeners to proclaim a change in hearing, reveal the complex relationship between distraction and contemplation in urban settings, including the possibility of refining sensory perception and awareness. In addition, the discourses surrounding listening practices at Off Site—often framed in culturalist terms—evoked widely held yet problematic notions regarding the process of listening in Japan and the acceptance of noise in urban public space more generally.

Like muzak, *onkyō* is a decidedly urban form, and it is often spoken of as

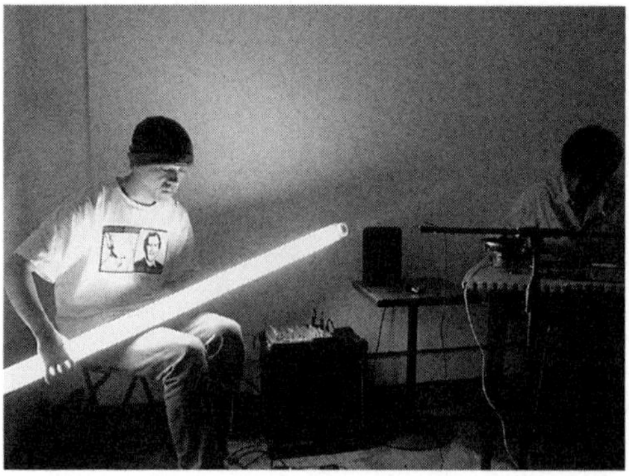

FIGURE 1.2 *left* Performance at Off Site. Itō Atsuhiro (left) performing on his self-designed instrument, the Optron; Tsunoda Toshiya (right). Photo by the author.

FIGURE 1.3 *right* Equipment set up before a performance at Off Site. Photo by the author.

being heavily informed by "Tokyo-esque qualities" (*Tokyo-teki jijō*). Opened in 2000 by artist and now full-time musician Itō Atsuhiro, Off Site operated as both a live performance space and a gallery, and came to serve as a rejection of Tokyo's long-standing tradition of rental galleries and music live houses in which performers pay rental fees. Itō designed Off Site in opposition to the structure of most art galleries, where spectators abruptly leave after viewing the artworks; instead, he chose a more inclusive gallery in which spectators were encouraged to linger at the café upstairs following music performances.

Located near Yoyogi station amid a cluster of postwar-style apartments on a small side street, Off Site's banal exterior—it was, in fact, a detached apartment—belied the "strange and curious" auditory happenings and events occurring inside the space.[12] Because Off Site was located in an apartment with poor sound insulation and in dramatically close proximity to neighbors, musicians (and by extension, their listeners) had to comply with rather unusual sound restrictions. All of the live performances had to be at such a low volume that outside sounds, such as traffic, pedestrians' footsteps, or traces of conversations, would frequently filter in and often commingle with or overpower the sounds of the performance itself. During Off Site's early days, a wall clock had to be removed because it became too disruptive during performances. The smallness of the music and its subtle shifts in development thus became one of *onkyō*'s most discernible characteristics, and the form became aligned with Off Site in media coverage and public discourse, including Tokyo's regularly occurring public dialogues or

FIGURE 1.4 *left* Sign outside Off Site from adjoining street. Photo by the author.

FIGURE 1.5 *right* Sign directly in front of ground-level entrance to Off Site. Photo by the author.

roundtables (known as *taidan* or *zadankai*). The owner Itō persistently denied a causal relation between the space itself and the emergence of the musical form. Instead, he emphasized the accidental eventfulness (*dekigoto*) that permeated performances at Off Site—such eventfulness refers to unexpected sounds from the outside (*soto no oto*) entering the performance space, as well as unexpected reactions and responses from the listeners.[13] These later interruptions, I will show, were not always viewed favorably by musicians.

ONKYŌ AND THE SENSES

Through its inextricable linkages with Off Site, *onkyō* gradually became known as a form that takes up background and ambient sounds as co-constitutive with its reception. Because many listeners spoke of a "change in hearing" (*mimi ga kawatta*) as an effect of Off Site's performance environment, it is important to examine the sensory and spatial ideologies surrounding the space itself. What exactly is at stake in the listeners' claims to have honed or regained their senses? Off Site itself provoked intersensorial responses in the listeners. Steven Connor points out that perhaps, paradoxically, the more we focus on one sense in particular, "the more likely it is that synaesthesic spillings and minglings may occur." He notes that "to look intently may be to long to grasp and consume; to be surrounded by sound is to be touched or moved by it" (2004, 153). Rejecting the pay-to-play model, Off Site's experimental philosophy was engendered, on some level, by the unexpected material and acoustic dimensions of the space itself. Off Site's deliberately minimalist architecture and décor—with its blank white walls and concrete floors—invoked the "white cube" aesthetic that Brian O'Doherty referred to as the guiding aesthetic principle behind gallery spaces. He explains the connections between the architectural structure of the church and that of the art gallery: "A gallery is constructed along laws as rigorous as those for building a medieval church. The outside world must not come in, so windows are usually sealed off. Walls are painted white. The ceiling becomes the source of light. The wooden floor is polished so that you click along clinically, or carpeted so that you pad soundlessly, resting the feet while the eyes have at the wall. The art is free, as the saying used to go, 'to take on its own life'" (1986, 15).

The notion of a gallery wherein art being exhibited operates autonomously by taking on a life of its own refers to the disciplinary ideology undergirding the model of the white cube. As Jim Drobnick argues, by shutting out and sanitizing the space from light and smell, thus maintaining an aura of neutrality and disconnectedness from the social and economic sphere, galleries operate as "single-sense" structures designed to "compartmentalize, isolate, and train perception" (Barbara Kirshenblatt-Gimblett cited in Drobnick 2004, 267). Itō designed the ground-level performance space and gallery in the classic form of the white-cube gallery: a windowless confined space with blank white walls and concrete floors. Yet while the space itself let in no outside light, thus evoking a sense of visual neutrality, sounds from the outside frequently filtered into the performance space, one of Off Site's unique characteristics embraced by musicians and listeners.[14]

According to Itō, listeners at Off Site understood the minimal sound conditions of the space and did not experience the live environment as one of restriction. Yet many listeners and musicians I spoke with frequently expressed notions of tension (*kinchōkan*) when speaking about their experience at live performances. One listener (and musician) explained the inherent tension in the space itself as imparting a sense of pressure (*atsuryoku*), noting, "It's narrow and cramped, and completely white." He made sure to add that this was not necessarily a negative sense of pressure, but rather a productive and challenging one. Others I spoke with echoed these sensations of tension, in part due to Off Site's cramped surroundings, in which listeners sit immobile in uncomfortable chairs, often awkwardly close to other listeners. This tension was palpable prior to most performances, as audience members would often sit quietly in their chairs waiting for the concert to begin, in a state of suspended silence. O'Doherty's comparison of churches to modernist galleries, where "one does not speak in a normal voice; one does not laugh, eat, drink, lie down, or sleep" (1986, 10), evokes sensations similar to those of listeners silently waiting in Off Site. The palpable tension in the performance space was frequently punctuated by unexpected sound events (*dekigoto*), referred to by Itō as *hapuningu* (happenings). Such events revealed not only the porous borders between the inside and outside of the performance space, but also the arbitrary divide between "music" and "sound." As Itō explained, "Sometimes during extremely quiet performances, you might hear the sound of a squirrel outside, or heavy rain, that would overpower the performance, to the point where you couldn't even hear the performance." The addition of such ambient city sounds was complicated by the concomitant presence of human sounds—an audience member falling asleep and snoring during the performance, for example—which were not always welcomed by the musicians. Such sounds coexisted uneasily in the live performance space.

Although the musicians desired contemplative listening, the (often barely audible) performances occurred alongside outside ambient sounds filtering into the space, passing in and out of the listeners' consciousness. This begs the question: Were Off Site listeners really listening to the music, or just simply listening? Here, focused, contemplative listening unwittingly converged with the sonic distractions of the tactile urban landscape. If audiences were unable to concentrate on the performance, whether due to the musicians' use of silences or the music becoming indistinguishable from the intermittent non-intentional sounds—a watch ticking, the air conditioner humming, outside traffic—did this not ultimately affect their ability to concentrate on the music? I am reminded of Jonathan Crary's notion of modern distraction as an effect of attentiveness: the more you pay attention, the more distracted you become (1999, 49). Some of the musicians and listeners I interviewed were cautiously open to the intrusion of such ambient city sounds seeping into Off Site's performance space. For them, however, the notion of outside sounds was a somewhat restricted category. It did not include the sounds originating from the audience's bodies, such as those who drew attention to themselves by breathing heavily, snoring, or accidentally dropping a beer bottle. Indeed, both listeners and musicians at Off Site drew distinctions between acceptable and unacceptable sounds. As one musician explained, "The sounds of cars or air conditioners during a performance are fine, but not people's voices."[15] There was an implicit sense among musicians, then, that human sounds (snoring, talking, dropping a beer bottle, and so on) threatened (aesthetically and practically) the concentrated listening that was such a crucial component of *onkyō*. Ambient sounds of the city coming from outside Off Site, however, were seen as acceptable. *Onkyō* could exist alongside the sounds of an air conditioner, but sounds such as voices or snoring carried the weight of human agency, which were ultimately perceived as intrusions.

Because of these non-intentional forms of ambient sounds at Off Site, many listeners and musicians felt their sense of hearing had dramatically changed over the course of the space's five-year existence. Some listeners described the process of listening to *onkyō* as a test of endurance or perseverance, a sentiment that reflected characterizations of *onkyō* as a form of music requiring listeners to "strain their ears" (*mimi wo sumasu*; literally, "to [clear] one's ears"). One listener explained his listening techniques during performances by a musician known for extremely abstract performances with extensive use of silences: "In Sugimoto's performances, there might be ten minutes of silence and you have to really be patient (*gaman suru*), but of course, during those times, I would start to focus more on listening to the outside sounds. And because of that, my sense of hearing has really changed." For this listener, the distraction of ambient sounds paradoxically strengthened his attention toward the musical performance, ultimately refining his aural sensibilities. Another regular listener explained how she noticed that her hearing had changed dramatically after the venue closed down: "What I realized afterwards, was that after going to Off Site, my hearing improved greatly. I was able to hear a lot of different things. I think that was a result of going to Off Site and hearing such small sounds and the kinds of echoes of the space." Such perceptual changes, however, were intertwined with awkward tension that was especially heightened during sparsely attended performances. The possibility or threat of such unwanted sounds, as well as the expectation of their irruption, was heightened in the space's subtly tense environment.

Soon after Off Site closed in 2005, one of *onkyō*'s most prominent performers, Otomo Yoshihide, explained, "If it was five years ago, I would have never heard the subtle sounds of which I'm now constantly aware." Similarly, a regular listener at Off Site recounted how his hearing had radically shifted due to attending performances at Off Site, and said that he is now able to hear much more delicate sounds (*sensai no oto*). He explained: "Before, I didn't appreciate sounds that had no connection to 'music,' such as a clock ticking. At the very least, I didn't understand the condition of these sounds. Even now, however, I still hate violent sounds, such as people's voices, especially in patronizing tones, as well as the announcer's voice inside the subway.[16] Aside from these sounds, I'm not bothered by any other sounds."

SERIOUS LISTENING

The process of listening in contemporary Japan is typically articulated in a specific cultural framework in which listeners are often problematically framed as "serious" (*majime*), prone to attentiveness and heightened studiousness toward the music itself.[17] This discourse tended to characterize discussions about listening practices at Off Site. Looking back, musician Otomo Yoshihide noted that Off Site's first few years of operation were its most creative. During these early years, *onkyō* had not yet become a fixed, marketable category or genre. Even more significantly, explained Otomo, it was unclear—for the performers as well as listeners—how one should listen to *onkyō*: "During the early period, the audience had no idea how to listen to this type of music. By the end of Off Site's operation, however, there was a sense of 'how to listen at Off Site,' which became well known and the style itself became somewhat fixed." Otomo also emphasized that the sound itself shifted over the course of Off Site's existence. Indeed, many listeners I interviewed observed that performances had actually become quieter as time went on, as the expectations of what an Off Site performance entailed became more defined, as Otomo noted.[18]

Jonathan Crary writes that the process of maintaining attention implies "disengagement from a broader field of attraction, whether visual or auditory, for the sake of isolating or focusing on a reduced number of stimuli" (1999, 1). Listening at Off Site, however, both reinforces and transcends this model of disengagement. For instance, some listeners made it a point to close their eyes during the performance because, as one listener told me, it was easier to concentrate on the "incredibly delicate, small sounds" if one's eyes were closed. If one's eyes are open, she explained, the performance is not as interesting, and it becomes difficult to focus on the restrained sounds of the performance. Closing the eyes, however, also evokes the sensation of having a dream, as she explained, which was made more surreal by the existence of external factors such as outside sounds. In this sense, the surreal and distracted quality of dreams was evoked by the audience members' act of closing their eyes. Ultimately, while they tried to sequester themselves from visual input, they could not disengage from surrounding sounds. Their auditory attention was repeatedly challenged by events occurring within the space—an audience member snoring, moving a chair, or arriving late, for example—as well as outside sounds penetrating the space.

Many listeners I spoke with not only were listening closely to the music being performed and the outside sounds, but also were aware of the acoustic dreamlike effect of such sounds on the walls of the performance space. One listener commented on the perception of sound within the architectural space: "I'm really listening, but there's a kind of pressure, you know? I'm still not sure what that is. As I'm listening, though, the sounds become kind of overturned (*hikkurikaeru*) and confused. For example, sometimes I can hear the sounds of the door, a chair, or a bike from outside all come to the front of the space. If I had to compare this effect to an artist, I would think of M. C. Escher. Sounds from the middle of the space, the back, and front shift and become reversed. That kind of thing happens a lot at Off Site."

ONKYŌ IS NOT AMBIENT MUSIC

Despite the aesthetic value placed on environmental sounds intertwining and overlapping with the sounds of the performance, *onkyō* was never intended to serve as background music, or music to be ignored, in the vein of "ambient music." Instead, the act of concentration (*shūchū suru*) in listening became the perceptual focus desired by the performers. Most of Off Site's performers and audience members were well aware of the lineage of avant-garde performance and environmental sounds out of which Off Site had emerged. Such a trajectory includes not only John Cage and Fluxus, but also composer Brian Eno, whose notion of ambient music, as a form of music that encourages and supports multiple "levels of listening attention without enforcing one in particular," invokes the ambiguity of background music, regardless of whether it should be ignored or listened to. As he explains in his manifesto, "[Ambient Music] must be as ignorable as it is interesting" (2004, 97). Eno defines ambience as "an atmosphere, or a surrounding influence; a tint" (2004, 96). His notion of ambient music was in opposition to Muzak Corporation's use of background music to add "stimulus" to the environment, a technique that encourages "repetitive consumption" (Attali 1977, 111). Eno's manifesto notes, "Whereas conventional background music is produced by stripping away all sense of doubt and uncertainty (and thus all genuine interest) from the music, Ambient Music retains these qualities. And whereas their intention is to 'brighten' the environment by adding stimulus to it—Ambient Music is intended to induce a calm and a space to think" (Eno 2004, 97).

Interestingly, the concept of ambient music first emerged from a period of sickness and immobility for Eno. Having broken his leg, he recalls lying in the hospital bed listening to harp music at a low volume. Unable to increase the volume, Eno describes his attention shifting to the way in which the sounds of rain and traffic intertwined with the harp music. While Eno's notion of ambient music aesthetics as a form encouraging thought without enforcing a specific ideology of listening ("as ignorable as it is interesting") might appear to evoke a perceptual awareness similar to that experienced at Off Site, they remain conceptually distinct forms.

Itō maintained that the modes of listening fostered at Off Site—the reception of environmental sounds alongside musical sounds—were not ambient, but were perhaps culturally ingrained. He argued that in Japan, such listening is cultivated early in life, noting that most Japanese have aural memories from childhood. Off Site, according to Itō, somewhat reawakened these earlier modes of aural perception, learned early in life, but forgotten amid the din and roar of the urban soundscape.[19] This prompts the question: How has the practice of listening in Japan been articulated, and can we explain listening as a culturally specific faculty?

LISTENING AND THE ENVIRONMENT IN JAPAN

The sensorium in Japan has often been framed, both in and out of Japan, in terms that maintain a sense of cultural distinction between Japan and the West, projecting these two spatial and legal entities as separate cultural objects.[20] Popular explanations for Japanese modes of listening, for example, often proclaim a heightened sensitivity and perceptual awareness toward certain sounds, often associated with nature, that are not valued or recognized by non-Japanese as meaningful. Many of my informants similarly argued for a cultural essentialist explanation of Japan's soundscape, often claiming that the very notion of noise in Japan differs radically from notions of noise in the West. During my interview with a musician and regular Off Site listener, he explained, "Japanese people don't make a distinction between a particular noise being seen as 'music' versus interpreting other noises as annoying [*jama*]." Linguist Tsunoda Tadanobu substantiates this culturalist view, arguing that the Japanese are linguistically and neurologically distinct from Westerners in that they have a "natural" auditory predisposition toward hearing "noise" (chirping of cicadas, for example) as music, or meaningful sound (1985).[21] Such a perspective continues to circulate within Japan, and it was often cited in explanations for modes of listening at Off Site. These discourses also emerged in my interviews with some sound designers in Tokyo, who drew on culturally recognized aesthetic tropes in their background music programming. The notion of a "natural" auditory awareness in Japan toward sounds that might be categorized as noise (such as insects buzzing, the tinkling of wind chimes, and so on) extends to discourses about traditional Japanese music as well. Extramusical sounds, such as the sympathetic buzzing vibrations of the *shamisen* strings, a phenomenon known as *sawari*, is often cited as an example of the acceptance toward noise as music within traditional Japanese musical aesthetics (Galliano 2002, 7).

Some musicians with whom I spoke about Tokyo's auditory environment confirmed the idea of a uniquely "Japanese" perceptual awareness of sound, including nature sounds. I interviewed a sound artist, for instance, who detected a shift in the quality of noise in Japan's public spaces across the past several decades. He lamented the loss of sensory engagement in contemporary Tokyo to what he framed as traditional Japanese sounds, such as wind chimes. For this artist, not only are such sounds being displaced by the constant din of urban noise, including muzak, but the sounds themselves no longer retain any aesthetic value. Such sounds tied to nature are now regarded as simply further examples of disruptive city noise.[22]

The hierarchical nature of such divisions—"good" sounds versus "bad" or "intrusive" sounds—as well as the nostalgic view toward sounds that are seen as disappearing evokes the work of Canadian composer and theorist R. Murray Schafer, whose theory of the soundscape was predicated on a distinction between hi-fi (rural) and lo-fi (urban) soundscapes. Schafer first coined the term soundscape in the 1970s to refer to the natural acoustic or sonic environment. His subsequent research in acoustic ecology grew out of a concern for preserving the natural sonic landscape, which he believed to be threatened by industrial modernization and noise pollution. Schafer's notion of soundscapes thus contained a strongly pedagogical dimension of listening to the world in a very particular way. This mode was also highly prescriptive. Schafer advocated "ear-cleaning" exercises as a means of documenting and attentively listening to sound in order to cope with a "busy, nervous society." He argued that the most important listening exercises are ones that "teach the listener to respect silence" (1977, 208). According to Schafer, ear-cleaning is a "systematic program for training the ears to listen more discriminately to sounds, particularly those of the environment" (1977, 272). Schafer's research on soundscapes, while important in its calls for engaging with our sonic environment, has long been criticized for privileging (or romanticizing) rural landscapes, while denigrating urban soundscapes.[23] As critics of Schafer have argued in defending urban soundscapes as meaningful sound, "a city wouldn't exist if it mirrored agrarian sonic space" (Arkette 2004, 162).[24]

Schafer's negative articulation of urban soundscapes as "lo-fi" suggests that such environments, brought forth by the Industrial Revolution, produce a sense of "sonic compression," thus creating barriers between urban listeners and their environment, which, he argues, ultimately weakens the listeners' sense of spatial identity (Arkette 2004, 162). One of the most emblematic sonic effects of the Industrial Revolution was the drone, or what Schafer refers to as the "flat line." Schafer observes that continuous sound, such as that produced by machines, is not found in nature and lacks temporal presence: "there is no sense of duration with the flat line in sound. It is suprabiological. We may speak of natural sounds as having biological existences. They are born, they flourish and they die. But the generator or the air-conditioner do not die; they receive transplants and live forever" (Schafer 1977, 78).

Similar to how artists and audience members alike explained to me that the ability to hear nature as meaningful sound was a uniquely Japanese cultural trait, a general acceptance of noise in public space is also often problematically explained as a "cultural inclination" (Dolan 2008, 663).[25] Tokyo's heavily saturated urban soundscape is the source of constant frustration and anxiety for many of Tokyo's residents, including many musicians and artists with whom I spoke. Their complaints often cite the constant noise—whether background music in cafés, train jingles, or politicians screeching their message via megaphones—as an example of "idiocy" or of "sickness," as philosopher Nakajima Yoshimichi has argued (1996).[26] Such responses in relation to Tokyo's urban soundscape complicate the idea that Tokyo residents are passive and unable to resist their aural environment, as previously noted.[27]

CONCLUSION

As a now defunct performance space on a small Tokyo side street, Off Site nevertheless articulated larger discourses over environmental sound, public noise, and the increasingly intrusive nature of Tokyo's soundscape, in particular, muzak. Despite (or perhaps because of) Tokyo's dense, cacophonous soundscape, listeners and musicians at Off Site claimed a physical transformation and refinement of their auditory perception. Such sensory awareness toward urban sound occurred, surprisingly, through detaching oneself from the urban environment. As both a rejection and a continuation of the white-cube gallery model, listeners at

Off Site were challenged by the controlled and restrained environment, which forbade smoking, talking, and, implicitly, sleeping. Because of *onkyō*'s extremely muted and sparse sonic and physical gestures, many listeners disengaged with the visual dimension of the performance, a process that allowed them to maintain focus and auditory attention. To concede such a shift in sensory perception, as many listeners and musicians admitted, implicitly denotes an initial loss and later regaining, or in this case, a reorientation of the senses. As C. Nadia Seremetakis writes, "To claim to have lost the senses, is to objectify them, to render them a discrete article that can be detached, then disposed of and eventually reappropriated" (1994, 123).

Yet, what allows for the initial "effacement" of hearing in Tokyo in the first place? In the case of Off Site, *onkyō* musicians call for active sensory engagement with Tokyo's non-intentional sounds, but simultaneously reject sounds coming from listeners themselves. Such a call is perhaps an implicit critique of the disciplinization of aurality in late capitalism, and could be interpreted as a critical reaction and response to the pervasive presence of muzak and urban noise in general within Japanese public space. As Steven Connor, Charles Hirschkind, and others have argued, navigating and taking pleasure in the city is crucially dependent on "subterranean forms of auditory knowledge and skill" (Hirschkind 2006, 21). Such auditory knowledge, or ear training, was cultivated by *onkyō* listeners in a manner perhaps recalling Schafer's notion of "ear-cleaning"—a necessary tactic for those living in "busy, nervous" societies.

Listeners occupied a *flâneur*-like position at Off Site; they were detached and present at the same time. The practice of concentrating on the silence ultimately made them more aware of the dense soundscape around them. Thus, rather than dismissing Tokyo's urban noise as a "sickness," listeners noticed that city sounds filtering into the performance space perhaps provoked new modes of listening. The city is no longer a "lo-fi" drone for Off Site listeners, but is made more alive and vibrant. For Taussig, via Benjamin, the advertisement "hits us between the eyes" (1991, 151), but perhaps urban sounds can achieve the same tactile effect. Is it possible to say that the sounds heard at Off Site had the tactile potential to hit the listeners *between the ears*?

The politics of listening that emerged at Off Site and their intersection with culturalist discourses concerning "Japanese" modes of listening reveal that sound is never a "neutral phenomenon" (Arkette 2004, 160). Sound is imbued with social and cultural meaning, its reception under continual negotiation and contestation. Sounds of Tokyo are hierarchically weighted, a sentiment that unfolded inside the blank walls of Off Site's performance space. Thus, despite the framing of *onkyō* as an uncontrolled and free form in which ambient, non-intentional sounds can and should be allowed to enter the boundaries of the performance environment, the very notion of ambient sounds was not all-inclusive in the case of Off Site. This "auditory discrimination"—in which the bodily presence of the audience is denied and/or downplayed in the face of the more valued ambient sounds of the cityscape—reveals the tenuous margins of the category of the urban soundscape as well as the impossible divide between music and nonmusic.

Perhaps we can explain *onkyō* listening as a critical response to what some have called "ubiquitous listening" (Kassabian 2001, 2013), here referring to the omnipresent nature of foreground music in capitalist environments. Background music, and urban noise in general, has become a naturalized part of the acoustic environment, seemingly sourceless and invisible, "posing instead as a quality of the environment" (2013, 10). This sheer ubiquity has necessarily provoked and informed the conditions of our listening; we can thus only listen in a distracted manner. As urban noise becomes more immediate and intrusive in public space, perhaps our critical engagement (or disengagement) with city sounds, as in the case of *onkyō*, will provide the conditions for new sensory engagement with urban soundscapes. The distracted listening that occurred at Off Site can be traced to other urban music venues in Tokyo, such as classical music cafés. The listening practices engendered by these leisure spaces illustrate the tension between contemplation and distraction as experienced by many Off Site listeners.

INTERLUDE I

City Noise and the Avant-Garde

Probably the first reference to noise as an explicit musical aesthetic was in Futurist painter Luigi Russolo's 1913 manifesto, "The Art of Noises," where he proclaimed the new noises of modern warfare to be the next step in the evolution of music toward "noise-sound." For Futurists, noise signified not only the urban atmosphere, but also the noises of war and industrialization. Founded by poet Filippo Marinetti in 1909, Italian Futurism emerged as a violent, pro-war, urban movement of poets, painters, sculptors, actors, and musicians. They gained notoriety through manifestos and public performances, which aimed at simultaneously provoking and insulting the spectator.[1] In his manifesto, Russolo argued that the category of music should be expanded to embrace all possible sounds, including noise. In order to sonically realize what he termed "the marvelous and tragic symphony of the noises of war" (cited in Tiekso 2016, 87), Russolo created new instruments, known as *intonarumori* (noise intoners), and gave the first public performance using them in 1914.[2] His embrace of noise foreshadowed early postwar musical developments such as musique concrète and John Cage's theories regarding silence and noise.

The founding manifesto of Futurism was first translated into Japanese and published by novelist Mori Ōgai in May 1909, just three months after its initial publication in the Parisian newspaper *Le Figaro*. The Japanese Futurist movement began in the winter of 1921, when poet Hirato Renkichi distributed copies of the first Japanese Futurist Manifesto to pedestrians in Tokyo's Hibiya Park (Sas 2005, 204). The manifesto began by declaring, "Trembling heart of the gods: the central activity of the human arises from the core of collective life. The city is a motor. Its core is dynamo-electric" (2005, 225). By the mid-1980s, in part

FIGURE INTERLUDE I.1
Futurism 2009 book cover (1986). Text at bottom reads, "Is it really coming? The 21st century."

due to the publicity generated by the 1986 International Futurism exhibition in Venice, Italy, a reassessment of the Futurist movement began to occur within the academy and among artists on an international scale. It was at this time that Italian Futurism experienced a boom amid Japan's expanding bubble economy in the 1980s.[3] During this time, many of the original manifestos were translated into Japanese, and musicians in particular began to reassess the impact of Futurism in the contemporary moment. Techno-pop musician Sakamoto Ryuichi—founding member of Yellow Magic Orchestra—released an album in 1986 titled *Futurist Boy* (*miraiha yarō*), which sampled an original recording of Marinetti reading his manifesto. Sakamoto also coauthored a 1986 book on Futurism with popular music scholar Hosokawa Shuhei, titled *Futurism 2009* (Figure 1), as a way of marking the hundredth anniversary of the publication of Futurism's founding manifesto.

Employing collage-like linkages and written in both Italian and Japanese, *Futurism 2009* is organized under three main topics: speed, noise, and movement. Within these sections, the book contains numerous subthemes. Above all, the book traces the flow of ideas from early twentieth-century Italian Futurists to theorists of the 1980s, noting in the preface that rather than architecture, construction, and the notion of the "original," the present postmodern moment speaks to ruins, deconstruction, and simulation. The book also features a dialogue between Felix Guattari and Sakamoto Ryuichi, in which they reflect on the contemporary notion of sampling vis-à-vis Italian Futurism.

The first intonarumori performance in Tokyo occurred in Shinjuku's Yasuda Seimei Hall on December 9, 1986. (Figure 2).[4] This performance was organized under the discourse of recreation (*saigen*) and featured some of Russolo's original Futurist compositions. The intonarumori replicas were created in the early 1980s under the guidance of then Tama Art University professor and music critic Akiyama Kuniharu, and they were played by his students during the performance.[5] Prior to his position as art professor at Tama University, Akiyama was perhaps best known in Japan as a music critic, and also for his involvement with the postwar avant-garde as an original member of the group Jikken Kōbō (Experimental Workshop). The 1986 concert also featured postwar avant-garde composer Takahashi Yuji on piano, as well as a public lecture on Futurist music.

The intonarumori from the 1986 concert remained unused, stored inside Tama Art University. However, they later resurfaced in Tokyo in 2002, in an intonarumori orchestra concert organized by one of my primary informants, the artist, musician, and owner of Off Site, Itō Atsuhiro. The concert was connected with Itō's solo exhibition at the university's museum, titled "Audible Vision 1913/2002," gesturing directly to the year of Russolo's manifesto. Having received his MFA at Tama Art University in the early 1990s, Itō was asked by the museum curators to organize an exhibition of his work. As Itō explained to me, he knew that the intonarumori replicas were sitting unused in the university's storeroom gathering dust, so he suggested using these instruments in a performance in conjunction with his exhibition. The orchestra featured six Tokyo-based experimental musicians who were all regular performers at Off Site. It may appear odd that some of the quietest musicians in Tokyo were chosen to perform on early twentieth-century instruments originally designed to reproduce the shocks and jolts of the urban soundscape. As Itō explained to me, however, it made perfect sense to take a "low-tech" instrument like the intonarumori that "disappeared at the peak of its fame," and to have it performed and interpreted by pioneering *onkyō* artists.

Itō was one of the orchestra performers himself, and during an interview we discussed his composition, titled "Russolo Phone." We also talked about Futurism more generally. Itō envisioned his piece as low-tech and machine-like, and instructed the orchestra performers to play mechanistically. It is against this backdrop that Itō's personal fascination with Russolo becomes clear. Maintaining

FIGURE INTERLUDE I.2

Flier for Italian Futurist intonarumori concert in Tokyo, 1986. Japanese text at top of page reads, "Recreated! Here! Now! The early twentieth-century 'Art of Noises.'" Underneath is an oft-cited image of the core group of Italian Futurists, taken in Paris, 1912. Left to right: Luigi Russolo, Carlo Carrà, Filippo Tommaso Marinetti, Umberto Boccioni, Gino Severini. Flier courtesy of Hosokawa Shuhei.

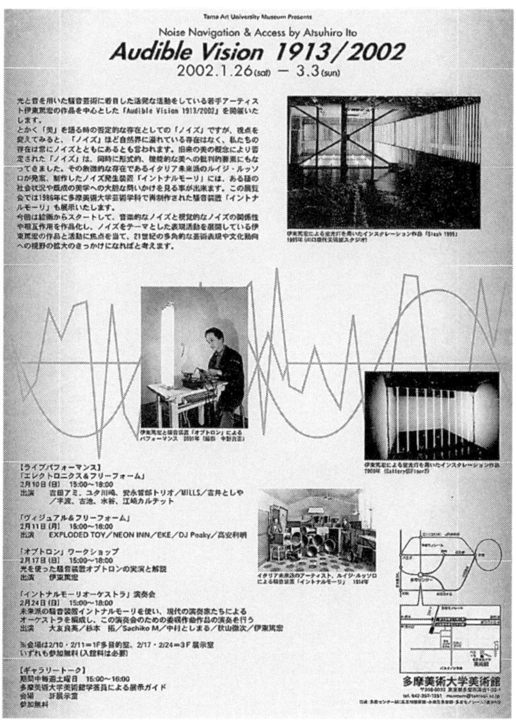

FIGURE INTERLUDE I.3

Itō Atsuhiro's exhibition flier, "Audible Vision 1913/2002." Image of Russolo and assistant Ugo Piatti in the laboratory, pictured at bottom of flier, suggests Russolo's influence on Itō's early art practice. Courtesy of Itō Atsuhiro.

little interest in Futurism as a movement—political, aesthetic, or otherwise—Itō was intrigued by Russolo, who, much like himself, approached the problem of sound and noise as a painter, rather than as a trained classical musician, or, for that matter, as a trained scientist.[6] According to Itō, Russolo was essentially a painter who loved music; however, given Futurism's eventual entanglements with fascism—and the fact that his musical inventions were consequently largely forgotten over time—Russolo stands as a figure that ultimately lost. Because he was not a musician, Russolo's experimentation was "foolishly bold," which appealed to Itō. His interpretation of Russolo's legacy echoes what others have suggested: although the instruments themselves were low-tech and crude, the ideas they provoked, particularly those of sampling and reproducing the sounds of the city, are ideas that continue to engage artists and musicians as well as shape their relationships with city soundscapes. As Itō stated while explaining his logic in choosing *onkyō* musicians to perform in a Futurist concert, "In what way should we listen to the city? How should we approach this problem? This theme continues up until the present moment. For that reason, I thought that city noise had a connection to the genre of *onkyō*."

TWO

"A Place Where Time Moves Slowly"

Analog Listening in the Music Café

HEAVY LISTENING IN THE CLASSICAL MUSIC CAFÉ

Two middle-aged male patrons are sitting in their chairs, coffee cups perched on the small table in front of them. Their posture is relaxed yet intent; their chins rest on their hands. They stare intently at the massive cabinet speakers and phonograph horns from which the music is booming as if the audio systems are live performers. Other patrons sit further away on the perimeter of the café; some doze off, some eat snacks they brought with them, some peruse the newspaper. Many patrons sit immediately in front of the speakers in order to directly experience the corporeal dimensions of the music and feel "their hairs stand on end" (Rothenbuhler and Peters 1997, 253). Listening is not just an aural experience, but also a bodily sensation. Unlike background music at most chain cafés, which is kept at a volume that deliberately marks it as "background" music, in the classical café, the music's intense volume fills up the room like a concert hall, demanding attention.

Tokyo's café culture is robust and dynamic. Like many major cities throughout the industrialized world, Tokyo contains numerous coffee shop chains, including Starbucks and Japanese chains such as Doutor. Alongside these chain cafés is a rich network of mom-and-pop coffee shops that are often framed as alternatives to chains such as Starbucks, which is valued mainly for its efficiency, not ambience.[1] Japan's food connoisseur community is cultivated in online forums and foodie review websites such as Tabelog (similar to Yelp), but it is also actively formed through print media, mainly through magazines devoted specifically to

FIGURE 2.1 *left* Heavy listening while facing speakers in Baroque, a classical music café in Tokyo that opened in 1974. Interior design is slightly different than in other long-standing classical music cafés in Tokyo, with 1970s era Formica tables and leather swivel chairs. Photo by the author.

FIGURE 2.2 *right* Regular listeners face speakers in Violon, a popular classical music café in western Tokyo known for warm sepia lighting and artful arrangement of phonograph speakers and tchotchkes lining walls and shelves. Front row of seats in sunken area facing the main speakers are believed to be the most acoustically ideal listening spots, according to the owner. Photo by the author.

reviewing cafés. One such magazine featuring lush photography, titled *Tokyo's Nostalgic Cafés*, recommends cafés that feature a blend of *wa* (Japanese notion of harmony) and retro atmospheres, in which the patrons can slowly relax and unwind (*yuttari kutsurogeru*).[2] The publication explicitly identifies this hybrid atmosphere with the postwar aesthetic, which is produced through the café's interior design, food, and significantly, its background music. It is not surprising that two of the cafés featured in a special issue devoted to nostalgia are classical music cafés.

In Japan, music cafés are more than just cafés that play music. They serve as public leisure spots for music fans and connoisseurs to listen to classical recordings (always vinyl) at a high volume, and they have strict rules of etiquette, including no talking and often no cell phones or laptops. Japan's contemporary classical music and jazz cafés first emerged in the pre–World War II era, when they served as a crucial site of leisure for the urban bourgeoisie. They flourished especially during the postwar period, when imported recordings and audio technologies were prohibitively expensive for the average Japanese urbanite. The antiquarian and leisurely ambience of these cafés stands in stark opposition to the hyperefficiency of chain coffee shops. It is precisely through analog sound, what I refer to as the café's "sonic patina," that café owners cultivate a nostalgic atmosphere described by one patron as "a place where time moves slowly." The listening that occurs in these cafés is unusual in that café patrons are "open-eared" and fully exposed to sound, unlike in cafés (both in Japan and globally) where customers often use headphones in order to create a private auditory cocoon, sealed off from their surrounding patrons as they do work and use the space efficiently. And yet, many patrons visit Japanese music cafés alone and, due to café rules, do not engage with other patrons while in the space.

This creates an intriguing listening environment. The auditory practices of classical music café customers belie certain assumptions about analog modes of listening, namely that the format demands focused attention on the exigencies of the medium itself (that is, having to turn the record over, and constant upkeep and maintenance of recordings). While many regular patrons listen to the amplified music in a concentrated manner (sometimes taking notes on the recording's jacket information), it is also the case that many people listen to the music as pleasant background sound, as one regular patron told me. It is ideal music to have in the background while reading, he noted. At the same time, analog listening demands a distinct temporal relationship to listening that is marked by slowness, relaxation, and contemplation. Significantly, relaxation is sought by moving *toward* sound-filled spaces such as music cafés.

In this chapter, I examine the techniques through which Japanese music cafés employ sound, specifically analog sound, to affectively manage their patrons. Classical music cafés and jazz cafés are historically tied to fostering expert, concentrated listening practices, yet they now serve as spaces for patrons to be surrounded by sound, often listening in a distracted manner while engaged in "nonproductive" leisure activities such as sleeping, reading, or daydreaming. Café customers find pleasure in the aimless and leisurely ambience through which sound serves a vital yet understudied role. The highly engineered ambience of the café affords multiple modes of listening for patrons, all of whom are affectively engaging (alongside their fellow patrons) with the amplified music that fills the space. Café regulars describe the attraction of the café in terms of its ambience, which is constituted by the space's antiquarian design, but also bolstered by the café's booming analog sound. I suggest that the cafés' continued appeal for an (admittedly small) audience is worth examining further. If most people are able to access and (often privately) consume digital music as they work, relax, and commute, what is the allure of cafés where patrons listen to analog recordings alongside other fellow patrons?

In researching this topic, I conducted participant-observation fieldwork at many of Tokyo's remaining classical music cafés, yet I focused my attention on three.[3] On one level, my ethnographic research was seemingly simple: hanging out for hours in the cafés and taking extensive notes provided me with rich ethnographic (and sensory) data. I found café owners to be interested in my research; I was able to talk extensively and informally with them, and I conducted many in-depth interviews. Meeting and talking with patrons, however, proved to be a daunting task, as most cafés have strict no-talking rules inside the space of the café.[4] As I came to know and become friendly with café owners, many of them offered to serve as intermediaries and introduced me to regular patrons. Having a formal introduction from the café master (as they are referred to in Japan) gave me legitimacy as a researcher and fellow café patron. In addition to participant-observation at cafés and interviews with owners and patrons, I collected print media about music cafés, including historical documents from their beginnings in the prewar era, as well as recent books and magazines in Japanese on café culture.

Grounded in a culture of high fidelity, phonographic listening favors the purported warmth and realism of analog recordings. The phonograph, radio, and tape recorders are quintessential examples of acousmatic sound technologies, defined by musique concrète composer Pierre Schaeffer as sound we hear without seeing its source. Although this disembodied sound was deeply unsettling for listeners in its early days, acousmatic, or background, listening is the predominant mode of auditory engagement with music today. Technologically mediated sounds revel in their ubiquity and "sourcelessness," becoming a part of the everyday aural environment (Kassabian 2013, 9–10). Café listeners desire a closeness and intimacy with the sound. In contrast to most cafés, where background music is projected from hidden speakers, the source of sound in classical music cafés is foregrounded and fetishized. The spatial design is arranged in order to bombard patrons with sound as seats are positioned to directly face the massive speakers. Patrons sit and listen (often alone) while facing the speakers. Some may sleep, read, take notes on the recording, write by hand in journals, stare into space, or work on laptops (although not all cafés allow personal computer use). Café patrons can make requests, although several café owners told me that many patrons do not take advantage of this option.[5] Customers are also able to bring their own LP records into the café, and the owner will play the recording on the café's sound system.

The modes of listening inside the café evoke what Anahid Kassabian refers to as "ubiquitous listening," the most common form of auditory engagement in the present moment. Brought about through the detachment of performance space and listening space made possible by the emergence of recording technologies, ubiquitous listening "blends into the environment, taking place without calling conscious attention to itself as an activity in itself" (Kassabian 2013, 10).

ENTERING THE MUSIC CAFÉ

One of the oldest of these cafés, Lion, lies in Shibuya, Tokyo's epicenter for consumer culture excess and nightlife, and home to Dōgenzaka, the famous narrow and hilly street closed to cars and crammed with "love hotels," where couples can rent rooms by the hour.[6] Heading up the narrow hill on the way to Lion, one climbs past a pet store, numerous sex shops, a hookah shop, a few fast-food chains, and several convenience stores, which dot Japan's urban landscape with such frequency their presence hardly seems noteworthy. The garish neon signs feel as if they become more insistent as sunset approaches. The background music projected at store entrances blasts passersby, yet it does not feel jarring or out of place here. Nestled within these shops devoted to immediacy and sensory pleasures—sex, tobacco, cute fuzzy animals, hamburgers, convenience stores, muzak—is café Lion. The street sign placed in front of the door reads "Built in 1926." It was rebuilt after World War II and the devastation of the Tokyo firebombings. Perhaps the owner wants to visually foreground to passersby the historical currency of this seemingly anachronistic mode of leisure. The nonproductivity of sitting silently and listening to classical music on vinyl while facing massive speakers for hours at a time seems like a potentially radical activity—a slap in the face of the noisy, obtrusive culture of instant gratification surrounding the café.

When I entered the café, the door creaked slightly as it gave way to my gentle pressure. Several patrons looked up as I sonically announced my entrance. The piano music that was playing was muted, so the creaking door felt all the more ostentatious, unshielded by high-volume symphonic music as I had hoped. Once fully inside, I was struck by the sensory disparity between the city energies that beckoned passersby and the muted café atmosphere, made more dramatic by the interior design. Thick analog crackles and pops lashed out at me. Dense plumes of smoke wove around the space as they slowly trickled toward the ceiling before dissipating into the yellowed rough walls. Frosted windows muted the city lights, sounds, and movements bubbling just outside the front entrance. Minimal light entered the interior space, a feature in the design theme for most classical music cafés. Inside the café, time seemed to slow down. This was not entirely due to the overtly retro furniture and yellowed walls heavy with patina. This felt like an altogether different kind of pleasure than the offerings that lay outside the café—convenience stores, hookah shops, sex shops, pet stores. People were sitting and listening intently while facing the speakers.

The servers at Lion, mostly college-aged men and women wearing white button-down Oxford shirts and black pants, were reminiscent of ushers at a theater or live music performance. Once I took my seat on the second floor, a young waiter silently approached me—eschewing the standard customer-service phrases that would normally accompany this transaction—and gave me a small glass of ice water, the sparse drink menu, and a copy of the music program for the day. I whispered my order to him (coffee with milk), and he nodded back silently. The woman next to me was sitting comfortably, leaning back in the chair facing the speakers and reading a paperback, seemingly oblivious to the music surrounding her. The middle-aged woman on the other side of her sat a bit more rigidly, as if in a state of suspended engagement, waiting for the performance to begin. She stared toward the speakers while taking occasional sips of her coffee as the recording blared out at us. Holding a handkerchief in one hand, she tapped her feet gently with the music's pulse, as if at a live concert. As I left Lion later that day, the city noises felt jarring after several hours in this atmosphere devoted to mediated listening and nonproductivity. At the same time, I also felt more attuned to and aware of the city sounds, something that listeners from different musical contexts told me.

Contemporary classical music cafés are suffused with discourses of nostalgia, disappearance, and the creation of ambience. They are intensely sensorial experiences born of the everyday life of the city—the rich scent of coffee with brandy and high-volume analog recordings heighten the patrons' olfactory and auditory senses (Hosokawa 2014, 112). Cafés where patrons can privately listen to analog recordings in a public space with other solo patrons would seem to be an outdated and strange concept today. Many of the owners I interviewed remarked on the anachronistic feel of classical music cafés in a world where music consumption is immediately accessible and listening habits are mobile, digital, private, and ubiquitous. Their comments almost reflected a sense of surprise, as if to say, "These cafés should not still be in business!" Some owners spoke of a desire to uphold and maintain music café culture, so it does not "die off," as one owner ruefully told me.

Referred to in Japanese as *meikyoku kissaten*, classical cafés and their analogue, jazz cafés (*jazu kissaten*), share a common history that is predicated on concentrated listening, emerging audio technologies, and collective listening practices.[7] Hosokawa Shuhei defines the jazz café as characterized by "the use of audio technology, imported music and drink, intensive listening, and bodily fixation" (2014, 110). Although the music genre differs between the two types of cafés, classical music cafés also fit this description. In both types, the owner is referred to by the English term *master* (*maasutaa*), or *madam* (or *mama*) if the owner is female. Regarding other Japanese terms that indicate mastery, Merry White points out that the Japanese word for teacher (*sensei*), translated as "the one who goes before," is typically used for traditional arts and is not appropriate for café owners, who are tasked with managing the sensory atmosphere of the café, maintaining customer service, and ensuring the quality of the coffee itself (2012, 66). In the case of music cafés, the master is positioned as the music curator and expert in charge of engineering the café's overall sensory ambience. The role of the café master or madam shares characteristics with the techniques of Tokyo's background music sound designers, who brand public space with programmed music or ambient sound, as I discuss in the following chapter.

Unlike the authoritative figure of the *sensei* in the traditional arts world, the café master is key to ensuring that the café is a success. In the case of music cafés, this affective management is a highly sensorial process that must draw in new patrons and also tailor the ambience to the tastes of the regulars. Café regulars often choose their venue based on the master's curation, referring not only to the aesthetic design of the interior, but more importantly to his or her music programming and knowledge of classical music. Patrons who do not own turntables or are unable to play loud music in their apartments are encouraged to bring in their own vinyl recordings to be played in the café. Café regulars often develop powerful bonds with the café master or madam through repeat visits. The relationship between owner and patron is strengthened by sound, here employed to manage the patrons' experience.

Regulars often view each café's music collection as unique. For some, it is the individual café's appeal. For example, at Mignon, a chic music café in the western Tokyo neighborhood of Ogikubo, the original owner and mama-san (who passed away in 2005 at age ninety-four) often comes up in conversation. The music catalogue is surprisingly extensive and unorthodox for a classical café, and both the current owner and the regular patrons are often quick to point out (with a sense of pride) how open-minded and eclectic the mama-san's musical tastes were. She liked "all types of classical music," I was told, including the music of avant-garde composers such as John Cage.

The name for classical cafés, *meikyoku*, translates literally as "famous piece" or "masterpiece." Commonly used during the prewar era to refer to Western classical masterpieces, *meikyoku* has come to serve as a synonym for Western, specifically European, classical music. The term was widely used in the 1930s for imported recording collections of masterpieces as well as print media (with detailed commentary) on such works.[8] These recordings and music criticism served a pedagogical function in legitimating classical music for the intelligentsia and also constructing the "masterpiece" itself. Listeners could rely on such imported music and texts to introduce them to Western classical music. Tokyo's music cafés serve a similar pedagogical function for their listeners; however, the very notion of the masterpiece (that is, what counts as "classical") is a contested category among café owners.

Music cafés are embedded within a complex history of urban leisure, sound reproduction technologies, and new forms of listening for city dwellers unable to afford imported recordings. With the increasing affordability of personal stereo systems, including the Walkman, during Japan's bubble economy of the 1980s, music cafés (both jazz and classical) decreased in popularity (Derschmidt 1998, 310). Some have argued that this decrease was partly due to a decline in serious intellectual interest in the music itself, as evidenced by the popular use of jazz and classical as muzak in Japan's retail and leisure spaces.

"STARBUCKS IS NOT A *KISSATEN*": PATINA AS NOSTALGIA

Once I interviewed a café madam who told me that she wanted to create the kind of place that she would like to visit. Starbucks or Doutor are "fine," she told me, but her café was special, and she wanted to preserve this kind of space. For many patrons, the appeal of visiting classical music cafés lies in the ambience and design of the café itself. In addition to the music selection, regulars frequently cite a café's atmosphere (*funiki*) as the reason they regularly attend. Many customers described their favorite café's ambience as being "just right" (*chōdo ii*) or "ideal" (*risō teki na*).

Of course, what makes a café's ambience "just right" depends on individual tastes. Many patrons told me that the sound system was crucial in creating the ideal ambience; one female customer explained that she liked the feeling of being surrounded by sound, here referring to the decorative placement of handmade speakers throughout the space. One regular told me he preferred his favorite café because it was the most "normal." For him, listeners at other music cafés either were too fanatical (*maniakku*) or were not actively listening at all, creating an atmosphere in which the classical music operated strictly as background music.[9] Others told me they preferred a particular café's ambience because it was less "strict" than other cafés. They could talk quietly with other customers (or the master or madam), or work freely on their laptops. The ambience of music cafés reflects a highly cultivated and managed aesthetic environment that is starkly opposed to that of chain store coffeehouses that are predicated on quick and convenient service, and feature harsh lighting and repetitive aesthetic design. The efficiency of chain coffee shops also extends to how patrons use the space, often as a private zone of productivity and labor, in which personal soundscaping (through the use of headphones) serves to create this privatized auditory space (Hagood 2011). Music cafés encourage differing notions of privacy by emphasizing leisure, distraction, and nonproductivity, all of which is facilitated through analog sound and equipment, without the use of personal audio technologies such as headphones.

In contrast to the efficient and standardized design of most chain coffee shops, music cafés are steeped in engineering a sense of patina, "the visible surface of a temporal depth" (Diaconu 2006, 132). Patina may refer to worn-down banisters or stairs, walls yellowed from years of cigarette smoke, or an antique mirror whose surface has turned green. As Madalina Diaconu explains, patina is a visual emblem that "appears when a surface is repeatedly touched over a long interval of time" (2006, 132). Yet it is also a highly dynamic and embodied process that is built on human interaction, thus evoking the haptic qualities of patina itself. "Patina stands for the ongoing process of the production of a lived space through the physical interaction between people and architecture" (Diaconu et al. 2011, 29).

FIGURE 2.3 Front entrance of classical music café Violon. Cozy seating area in front of the café is sometimes used as a smoking area, after the master recently barred patrons from smoking inside. Interestingly, this change in café rules was not due to health concerns, but instead, was intended to alleviate negative effects of smoke on the analog equipment. Photo by the author.

Stepping into a classical music café evokes an earlier period in Japanese history that is sometimes referred to as "Showa retro." The Showa era was the reign of Emperor Hirohito from 1926 to 1989, and Showa retro generally denotes the postwar era, roughly from the 1950s to the 1970s. Classical music cafés, like many other *kissaten*, are known for their Showa retro feel. Yet this putatively "Japanese" sense of retro (evoking the aesthetics of postwar Japan) is often entangled with European aesthetics.[10] Many cafés are given European names, for instance. In one notable case, the storefront sign is written in French. Windows are typically opaque (or, no windows at all), thus shutting out external light and visually and aurally orienting focus toward the analog sonic atmosphere unfolding inside the café. Most cafés favor dark or neutral colors and small wooden tables. Walls and shelves are often crammed with a hodgepodge collection of nostalgic objects—dried flowers, framed paintings, antique lamps, matchbooks from Tokyo's classical music cafés, European figurines, and tube amps on display, among other objects. Even the furniture evokes the nostalgic feeling of the postwar era. Some cafés, for example, feature high-backed crushed red velvet chairs, often with white lace doilies (similar to those used in the headrests of Japan's cabs) or white covers draped over the backs of chairs.

The owner of Marguerite, a classical music café in western Tokyo, told me that the height of the chairs was adequate when the café first opened in 1955.[11] However, as people have grown taller since the postwar era, the chairs are too small for most young people today. The unusually low chairs and high wooden tables, out of sync with one another and the tight spaces between tables, force patrons to be aware of their bodily comportment in navigating the space inside the café. They have to conscientiously avoid collisions with other patrons' tables, as well as try to stay quiet.

A regular patron once explained the appeal of his favorite music café to me by stating that its atmosphere (*funiki*) evokes a feeling of time moving slowly. When in the café, patrons are encouraged to relax, do nothing, and feel leisurely. This is a concise evocation of the phenomenon of patina, "which converts time into a positive aesthetic agent" (Diaconu et al., 2011, 29). This same patron told me that the café's ambience felt "nostalgic," even though he was too young to have experienced 1950s culture firsthand. Jordan Sand describes the consumption of 1950s nostalgia in 1980s economic-bubble Japan, noting that it reflects a more "generalized condition of ambivalence about capitalist modernity in Japan" (2006, 86). This feeling of nostalgia (or the pleasure of consuming nostalgia for nostalgia's sake) not only is due to the café's retro furniture and interior design, but is also

FIGURE 2.4 *left* Postwar-style velvet chairs in Marguerite. Wooden branchlike poles create privacy between tables. Photo by the author.

FIGURE 2.5 *right* Red velvet chairs in Mugi, a classical music café near Tokyo University. Carpeted floor dampens the sound and creates a feeling of textural and visual continuity between the velvet seating area and the floor.

produced and amplified by the analog *sound* in the café itself. Indeed, many café patrons are too young to have experienced the vinyl era directly, and yet, the affective experience of the music café as a space where "time moves slowly" is deeply bound up with nostalgic assumptions about the analog format more generally.[12]

SONIC PATINA AND TACTILITY

One regular patron I spoke with described the effect of hearing a gramophone recording for the first time in the music café and the shock of hearing the sonic historicity and traces embedded in the disc itself. "So this is what it used to sound like for people listening to music!" he told me in a voice that gestured to the astonishment he felt hearing this format for the first time.

The notion of patina can be deployed more broadly to describe objects for

FIGURE 2.6 Antique patina at Marguerite café. A female patron I interviewed described the atmosphere at this café as "ladylike" (*jyosei poi*). Photo by the author.

which the passage of time is instrumental for consolidating and elevating their aesthetic value. For example, patina can be used to describe wines, since their scent and taste grow over time. Even archaic words can be seen as embodying a sense of linguistic patina (Diaconu 2006, 133). I suggest that we can apply the notion of patina, as a necessarily haptic phenomenon, to the sensory experience of sound and listening inside the space of the music café. Music café listening is an embodied and collective process that accommodates multiple, often competing, auditory modes. Ubiquitous listening, the predominant form of listening in the present moment—that is, listening while engaged in other activities—occurs alongside concentrated listening. The patron's engagement with sound is consolidated under the café's cultivation and management of sonic patina, which is predicated on listening as a tactile and corporeal technique. Café owners exploit the tactility of sound through specific architectural techniques in order to enhance the listening experience. For example, the owner of Marguerite described how the café's unique interior design significantly affects the acoustic dimensions of the space. She pointed out that the rough and bumpy walls are ideal for sound to echo and reverberate throughout the café. The high ceilings also allow sound to resonate throughout the space, she informed me. These ideal acoustic conditions are experienced by customers seated at private listening tables, described by the owner as "private worlds." These private nooks are created through large bamboo-like structures that are placed between each table, so that patrons are separated from each other. Along similar lines, the owner of Violon modeled the café's interior design after the structure of a Viennese concert hall. With a lifelong interest in the relationship between acoustics and architecture, the master dramatically transformed the café's interior (originally a sushi shop) to create an ideal listening environment. He achieved this most notably by changing the floors from the standard Japanese style of solid cement (in the case of retail spaces) to wooden floors, ideal for reverberation.

The café atmosphere is generated through extensive staging, so that patrons experience sonic patina in a distinctly tactile manner. The haptic sense has often been described in terms of visuality, specifically in the medium of film. Laura Marks posits the notion of "haptic visuality" as distinct from optical visuality, which presumes a separation between the viewer and the object being viewed. Haptic visuality, she argues, is inherently tactile, "as if touching a film with one's eyes" (2000, 162). The "skin" (Diaconu et al. 2011, 28) on objects imbued with patina as a result of repeated touch is akin to the materiality, or "skin" of the film that gestures to the tactility of vision itself. The corporeal and relational (and thus inherently social) nature of patina resonates with Marks's notion of film as skin, which invokes the "effect of a work's circulation among different audiences, all of which mark it with their presence" (2000, xi).

Sonic patina draws on the inherently haptic quality of patina as an attribute of objects that is produced through repeated human touch. I conceive of sonic patina as an embodied and intersensorial aural experience critical to the production of the café's space that cannot be reduced to the ears alone. The warm and resonant analog textures of the recordings contrast uneasily with the café's creaky wooden floors, which puncture the hushed silence surrounding the amplified recordings blasting out of vintage speakers. The deep red velvet chairs feel plush, and yet their out-of-sync low height and faded color evoke a strong feeling of patina, the traces of which "save and record gestures, feelings, and memories" (Diaconu 2006, 134).

Sound is received as a necessarily tactile and bodily experience; it "literally touches our bodies, moves and vibrates the inner workings of our ears and echoes through our bodies in order for us to hear" (Coulthard 2012, 18). As the product of the violent collision of two objects, sound is also grounded in power relations. Steven Connor describes the tactile nature of sound as containing a sense of pathos and sadism. He writes that "sound is imagined in the same two-sided way as skin: both as that which touches and that which is touched, both as a medium through which we feel and as something that is itself subject to touching and assault" (2004, 163).[13]

During the middle of my lengthy visit to a music café in western Tokyo one day, three patrons in their thirties entered, two women and a man. They appeared to be co-workers. Based on their hesitation when entering the space and their surprise at seeing the cramped conditions (and being hit by the café's loud music), I guessed that they were new customers, perhaps unaware of the no-talking regulation (which is not actually written down anywhere at this particular café). They sat in the sunken area in the middle of the café and, oblivious of the other patrons who were completely silent, immediately began talking loudly and animatedly. Their voices sounded jarring against the crackly vinyl recordings booming throughout the space. Several customers glanced over, disturbed by how the loud conversation clashed with the room's acoustics. The madam came to take their order (iced tea and iced coffee), and I overheard her politely telling them to "speak more quietly." They apologized for their behavior, and quickly finished their drinks and exited, after which they sat outside the front doors, chatting and smoking.

SOUND ETIQUETTE AND LISTENING CONVENTIONS

Sound is under great scrutiny and highly managed by the master or madam inside the space of the café. While most cafés have restrictions on talking inside, non-intentional sounds have a curious place within the café's soundscape. For example, Merry White describes how Ryugetsudo café in Kyoto painstakingly tries to prevent such sounds: waitstaff wear felt shoes to dampen the sounds of movement; patrons who buy pastries from the bakery downstairs must remove the crinkly paper before entering the café. Some tables do not allow patrons to write; drinks are placed on a leather mat along with a wooden spoon so as to mute potentially disruptive sounds of clanking (2012, 62).

Despite the emphasis on focused, contemplative listening, which would seem to demand a quiet interior space, the floors of many cafés are extremely creaky, thus causing listeners to be constantly aware of their own bodily presence within the café. Regular patrons choose their seats carefully and often have a regular seat. As one customer explained to me, for some patrons the seat is chosen based on proximity to the speakers, but for others it is chosen so as to be the least disruptive to other patrons. This awareness of other patrons can extend to the question of taste and its impact on the café's overall listening environment. One regular patron told me that he will not make a request unless he is the only patron in the café, so as not to disturb the atmosphere if he were to choose music that other patrons do not find pleasurable.[14]

The owner of a wine bar around the corner from a music café told me of the time he entered the café to drop off fliers. He felt extremely intrusive, and felt compelled to walk as lightly as possible inside the café so as not to disrupt the "serious" listening environment. Many owners and staff gesture to the past when speaking of classical cafés, often pointing to apocryphal stories: "They used to be so strict, you couldn't even read a book; you were only allowed to listen!" Many point to a recent loosening of restrictions on the listeners, perhaps as a way to encourage new patrons as chain coffee shops rise in popularity.

Regardless of these subtle shifts in restrictions, most cafés adhere to certain listening rules. These are often enforced through the structure of the café itself. One café, Mugi, located near Tokyo University, announces such rules through its spatial layout. As patrons descend the stairs into the basement café, they are confronted with a choice to sit in the room on the left (no talking) or the right (talking allowed). I conducted an interview at Mugi with a Tokyo University professor, who was also a longtime classical music café patron. He explained

FIGURE 2.7 Sign outside Baroque café (upper right-hand corner) that reads, "Because this is a music appreciation spot, talking is prohibited." Photo by the author.

these distinctions to me, for there are no signs written indicating the difference in rules between the two spaces. Some cafés explicitly announce their rules through signs at the entrances indicating that talking or the use of cell phones is not allowed, or that smoking is not permitted.[15]

Another café does not allow the use of laptops because their presence "conflicts" with the overall ambience of classical music cafés and would be disruptive to other patrons. This rule is not written down anywhere; however, during my first visit to this particular café, the owner politely reprimanded me after I opened up my laptop to take notes. The conflict is that the material presence of a laptop, with its white screen, causes a visual distraction to other patrons, but the laptop is also heard as disruptive because of the sound of the keys clacking on the keyboard. The space is also one that is not necessarily a zone of work (as perhaps is true with chain coffee shops), but is rather a space of leisure, relaxation, and listening. A young female patron explained the appeal of her favorite café's listening environment: "You have patrons who are extremely knowledgeable about classical music who are listening intently, but then you have other patrons who are sleeping right next to them."

ANALOG LISTENING AND THE AUDIOPHILE

Interest in audio culture among middle-class men increased in the 1920s and 1930s, as electric record players and radios were imported into Japan amid a burgeoning market of specialist music magazines targeting boys and young men (Hosokawa and Matsuoka 2008, 41).[16] Japan's audiophile community strengthened dramatically during the postwar era as Japan rebuilt. This was primarily due to the emergence of Akihabara ("Electric Town"), a neighborhood in Tokyo that served as an "informal gathering place for black marketers, compradors of illegal outlets from American forces, buyers of electronic components and those who needed their radio repaired" (2008, 42). As the epicenter for audio manufacturers and electrical components, Akihabara gave rise to audio specialist subcultures (including hobbyists and tinkerers) that continue to this day (2008, 42). By the 1960s and 1970s, as Japan's economic strength rapidly increased, the middle class was able to buy domestic electronic products including stereo technologies, which were marketed to consumers as part of a "modern" lifestyle (2008, 42–43).

When the phonograph first emerged as a domestic audio technology for the middle class in the first two decades of the twentieth century, the act of listening to disembodied recorded music, especially alone, was initially viewed as a shocking experience. Mark Katz describes the shift in listening experiences that occurred in this moment. In the pre-phonograph era, he writes, listening was a collective cultural endeavor that accompanied significant events such as weddings, births, or deaths. With the advent of recorded music, however, the concept of solitary listening emerged. Although solitary listening eventually came to be seen as a socially acceptable (perhaps even ideal) listening practice, British writer Orlo Williams noted in 1923 that it would be "odd" or embarrassing to walk in on a friend listening to music alone (Katz 2004, 17).

Katz points out that private mobile listening is the predominant mode of auditory engagement in the present moment, facilitated by private audio technologies such as headphones or earbuds. There are, however, instances where listeners "misuse" these privatized technologies in ways that create a shared auditory experience. Tyler Bickford examines the media experiences of preteens who share earbuds with their peers, thus instrumentalizing "listening as a sociable practice in the nooks and crannies of everyday life at school" (2014, 350). This practice, Bickford suggests, is comparable to ubiquitous communicative practices children engage in during class, such as whispering or passing notes. In this way, listening serves "less to emphasize particular music than to forge connections in the background of group conversations or other activities" (2014, 337).[17]

The experience of analog listening in the music café complicates the standard logic of post-Fordist auditory experience as necessarily private, mobile, and inward-looking. It is also a mode of listening that demands a slower mode of engagement. However, café listeners coming together and listening collectively is not a communal auditory experience in the way it is for Bickford's earbud-sharing schoolchildren. Mediation and circulation are "assumed, naturalized, and backgrounded" for his young listeners, who are not turned toward the source of production (or an "original acoustic event") (Bickford 2014, 350). For listeners in cafés, on the other hand, the mediation itself, facilitated by the fetish of analog technology, is foregrounded and emphasized in their very listening space (2014, 30).

Media historian Lisa Gitelman has described the audience for the public phonograph parlors in the late nineteenth century, noting that people listened to records via ear tubes. Thus, a "public experience was in another sense, a profoundly private one" (2003, 63). As she explains, the phonograph medium itself "helped divide customers from one another even as it drew them into crowds and helped imagine them as communities" (2003, 64).

The listening community at music cafés comes together in unexpected ways. Despite the restriction on talking at many cafés, regular patrons told me how they became acquainted with other regulars. I recall one time at a music café where I was seated near several patrons who were all listening attentively to the recorded sound. The man seated nearest me was sitting upright in his chair facing the speakers, almost leaning forward to take in the booming sound. I felt pangs of guilt when I briefly checked my phone for email messages concerning an upcoming interview. Although we were sitting separately and not interacting at all, the listening atmosphere at that moment evoked the atmosphere of a live performance that we were collectively experiencing.[18] Checking my phone felt like a rude disruption to the heightened auditory environment. Unlike most classical music cafés, Mignon allows talking, as long as it is at a low level. Cell phones, however, are prohibited. I spoke to several regular patrons at Mignon about their experiences, and they all cited the ability to talk with other patrons and, most important, to talk to the madam, as the café's main appeal.[19] I interviewed a white-collar office worker, referred to as "salaryman" in Japanese, in his forties who has been a regular at Mignon since his high school days. He used to attend other cafés, but now he goes only to Mignon, and he specifically cited the ability to sit at the bar counter and talk freely with the madam as part of the café's charm.[20] He told me that he felt that Mignon evoked a strong sense of belonging (*kizoku ishiki*) among regular patrons and compared it to regulars at a bar.

Similarly, I chatted with a middle-aged couple who have been attending Mignon since its early days with the original madam. Surprisingly, they told me that the relationship they have cultivated with the café madam and other regulars is the primary draw for them, rather than the music per se.[21] The husband bemoaned the shift among young people in Japan today who prefer going to chain cafés (he used the English term *kyafe*, implicitly invoking chain coffee shops like Starbucks or Doutor), rather than *kissaten*, which are slowly disappearing as an institution. For him, the chain café engenders solitary, antisocial behavior, whereas the *kissaten* is ultimately a social space.

ANALOG RECORDINGS IN THE MUSIC CAFÉ: STORING, COLLECTING, AND LISTENING

Sonic patina is produced by the analog format favored by music cafés. With a single exception, classical music cafés in Tokyo play only analog recordings.[22] I spoke with the owner of Violon about this. He very adamantly told me that he plays analog recordings to be true to the history of classical music cafés, as cafés originally only played analog recordings. He explained that his choice to play only analog recordings was not based on viewing them as aesthetically superior to compact discs as an expression of audiophilic purity. As sensory objects that operate as a "repository of history, sentiment, and memory" (Peters and Rothenbuhler 1997, 255), the café's LPs serve a critical role in defining the café's antiquarian sonic ambience. While the category of "classical music" might appear to be self-evident, all music cafés offer different catalogues and define "classical" differently, especially in terms of when the classical period ends. One café, for example, holds recordings in its collection up until Schubert; Romantic music is the least featured genre in the café's collection, the owner told me.

The massive display of recordings and patrons' devotion to listening also conjures up images of the audiophile—a music listener who seeks high fidelity sound through the accumulation of specialized high-end audio equipment. The audiophile's preferred listening environment evokes the music café's spatial design and architecture, which is intended to create the ideal acoustic space. For example, as Eric Rawson explains, the audiophile's listening environment is "usually a space separated, like a bedroom, from the rest of the living quarters to ensure privacy. Custom-built or modified from the existing structure, the listening room is designed to correct acoustic anomalies and permit clear transmission of the musical signal. Typically, the walls and ceilings have been altered to prevent parallel planes, the reflections from which cause standing sound waves and a loss of sonic information" (2006, 205).

Audiophiles are also known as expert listeners, based on years of accumulated knowledge and specialized training. The owner of Violon is held in high esteem not only among café patrons, but also among musicians in Tokyo for his extensive knowledge of vintage audio equipment. Always dressed in a white button-down and black pants, the master's everyday sartorial formality is belied by his kindness and openness to his patrons. His mastery of audio technologies is legendary; he makes and repairs the tube amp speakers in the café, which is one of only two cafés in Tokyo to feature this old technology. He is also fascinated by the relationship between sound and architecture, largely a result of having lived in Vienna as a young man. The master points to his audiophile expertise as a central reason that his café has remained in business. Unlike most café owners, he builds the café's tube amp speakers himself, thus saving a significant amount of money.

In addition to cutting costs on audio system maintenance (an expense that nearly all cafés must deal with), Violon's nightly live performances bring in a steady flow of patrons who are generally different from the daytime clientele.[23] Other cafés that do not offer nightly performances must charge exorbitant prices for drinks or snacks in the face of a dwindling clientele. One elderly female café owner told me that her café is always "in the red" because fewer people now go to classical music cafés, and because the neighborhood itself (Kichijoji, in western Tokyo) is changing and becoming "less intellectual."

As a medium favored by audiophiles for a truer (high fidelity) representation of the original performance, the analog format is undergoing a renaissance among music collectors globally, despite claims that the LP is a dead medium. Japan in particular is known as a haven for record collectors, particularly those searching for original and often obscure LP recordings, especially jazz.[24] Audiophile culture, including record collecting and audio technology fetishism, can be traced back to the 1920s, when Japan saw a concomitant expansion of the middle class and mass consumption. Internationalist policies at the time created consumer interest in Western culture more generally, and Western classical music came to be seen as emblematic of refined taste and high culture. Because there were very few live classical music performances in Japan, recordings "often provided the only audible access to Western music pieces" for Japan's educated middle

FIGURE 2.8 Street sign for Mignon café emphasizing analog stance. Photo by the author.

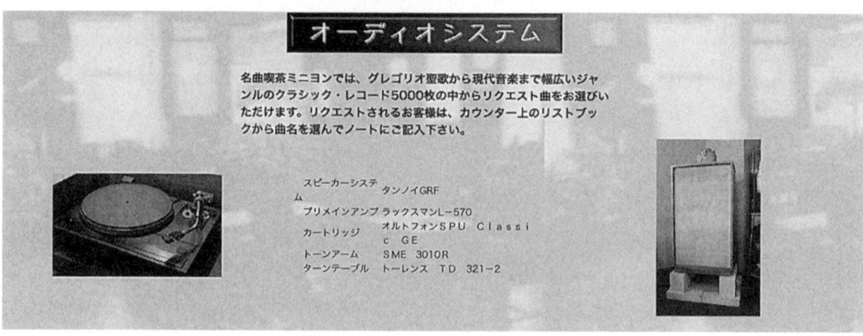

FIGURE 2.9 Mignon website page on audio systems, featuring photos of speakers and turntable system.

class (Hosokawa and Matsuoka 2004, 152). Music cafés were bound up with this burgeoning culture by offering music connoisseurs auditory access to expensive imported recordings as well as high-end audio equipment. Recordings brought Europe aurally to Japan, so to speak, as the interior design of Japanese cafés simulated European counterparts both spatially and visually.

MUSIC EQUIPMENT AND THE FETISHIZING OF ANALOG TECHNOLOGY

In prewar Japan, music cafés featured high-end audio equipment in order to provide the clearest and most "faithful" sound for their patrons, especially audiophiles. Audio technologies continue to be foregrounded by contemporary cafés through their online advertising, in-store marketing, and spatial design. In the music cafés of the prewar era, Hosokawa Shuhei notes, massive phonographs were spotlighted in the space as if on "an altar" (2014, 117). The near-spiritual appreciation of audio technologies in music cafés was underscored by the fact that the brand names of high-end audio equipment were often advertised in cafés themselves. This information interested music connoisseurs especially. In both jazz and classical music cafés, the music's intense volume was overwhelming for the listener. This remains the case even today. As many café owners and patrons explained to me, people find pleasure in listening to music in cafés at a high volume, something they are not able to experience in their home environment, since many in Tokyo live in close proximity to neighbors.

Stereo equipment and speaker brands are often foregrounded on cafés' websites and in the spatial design of cafés themselves. The website for café Mignon, for instance, features a section detailing its audio system with photos and descriptions of its stereo and speakers (Tannoy speakers). Similar to how restaurants or cafés might display images of featured items on their menus to entice patrons, Mignon advertises its stereo technologies. It also details its "wide" record collection of over five thousand albums, ranging from Gregorian chant to modern music (*gendai ongaku*). While each café's collection defines the temporal borders of classical music differently, the collection at Mignon is far more wide-ranging than most: most cafés, for example, do not include modern music by such postwar atonal composers as Varèse, Stockhausen, Xenaxis, and Cage. Notably, while most classical cafés feature only European composers, Mignon also includes Japanese composers in its collection.

Other cafés produce and circulate fetish objects that serve as advertisements

FIGURE 2.10 *left* Matchbook from Baroque café advertising audio equipment (front). Photo by the author.

FIGURE 2.11 *right* Matchbook from Baroque café advertising audio equipment (back). Photo by the author.

for their audio equipment. Opened in 1974, Café Baroque produces and distributes matchbooks that serve as marketing for the café (by including the address, telephone number, and business hours). These matchbooks are collectible objects, while also serving to emphasize the café's use of analog technologies. The front of the matchbook, which reads "Our Master's Voice," appears to be a play on the HMV gramophone company's name ("His Master's Voice"). The balding, bespectacled middle-aged man depicted on the matchbook sits reverentially before the gramophone in a formal tuxedo. Although it is unclear if he is Japanese, it is notable that he is sitting with his legs folded under him, a formal sitting position (*seiza* in Japanese) that is linked to the bodily posture for traditional art forms such as the tea ceremony. Strapped to his back are a violin and a *katana* (traditional Japanese sword). It is the gramophone horn, however, that holds our gaze. The listener (and musician) is lowering his head to the device, as he is confronted by the gramophone's sonic output, here represented by music notes dancing outward and upward. The matchbook image is clearly gesturing to the famous 1899 painting by Francis Barraud of a dog, Nipper, listening intently to a gramophone, his head cocked toward the sound flowing from the horn. Titled *His Master's Voice*, this painting became an iconic logo directly linked to the HMV brand.

The music recordings publicly display the café's music collections and define the aesthetic parameters of each café. In so doing, they legitimate the authority of the café master or madam. Shuhei Hosokawa notes that during the prewar

FIGURE 2.12 *left* Record collection at Mignon café. The café's catalogue includes over 5,000 albums. Patrons can make requests directly to the madam. The jacket of the album that is currently playing is placed on a stand above the countertop in view of patrons. Photo by the author.

FIGURE 2.13 *right* Record collection at Violon, one of the only cafés in Tokyo where patrons can physically browse through the recordings themselves. At other music cafés, patrons make requests through the (usually handwritten) catalogue. Photo by the author.

era, jazz cafés became a public space for record collectors to meet and listen collectively with other connoisseurs. The café's record collection was often partially listed in advertisements as a way to attract "maniac" patrons (2014, 117). Unlike digital formats such as compact discs, LPs "invite collecting" (Rothenbuhler and Peters 1997, 243). As Eric Rothenbuhler and John Durham Peters point out, it is not surprising that Theodor Adorno emphasized the fact that "stamps, photographs, and records are all kept in those 'herbaria of artificial life' called albums" (1997, 243). With the emergence of the phonograph, music could now be "possessed as a thing." The ostentatious display of the café's music archive points to the sense of fanaticism by which audiophiles (especially record collectors) have been categorized (and sometimes critiqued).

ENGINEERING COMFORT FOR THE LISTENER

One night I headed to Mignon, one of the cafés where I spent a great deal of time throughout my fieldwork. When I arrived a little before 7:00 p.m., I saw that the café was empty. After greeting the madam, I settled into my preferred seat, a cushioned bench against the wall that faced the bar counter. This seat gave me the best vantage point to observe other patrons as well as the madam selecting music from the massive record collection displayed on wall shelves behind the counter. While most classical music cafés actively shut out light and sounds from the street so as to retain visual and sonic emphasis on the analog sounds, Mignon is unusual in that its windows and doors are often left wide open, letting in sunlight and noises from the street: traffic, sirens, and dogs barking. This particular night was unusually breezy, and with every gust of wind, the front door dramatically flew open before gently shutting again, a movement that would be repeated incessantly throughout the night. Outside sounds and textures were filtering into the space and intermingling with the music with wild abandon.

At 8:15 p.m., I decided that I would make a request. I recalled the madam telling me the previous week that she had one Ravi Shankar album in the café's collection. Since I was still the only patron in the café, and this was a decidedly unorthodox choice of music for a classical music café (where classical implies European classical music only), I decided to take advantage of the emptiness of the space. Only moments later, and before I could get up from my seat to approach the counter and make the request, I heard the familiar drone of the tanpura resonating throughout the café.[25] I immediately recognized this as the Ravi Shankar recording she had mentioned to me. I recalled that I used to own this recording on cassette, back when I studied North Indian classical music. This was the legendary 1967 duet between Shankar and American violinist Yehudi Menuhin, titled "West Meets East." I initially felt shock at hearing the exact recording I was planning to request; she had predicted my request before I was able to make it. My initial surprise, however, quickly shifted to affective pleasure as I remembered the conversation I'd had with the madam the week before, when I told her about my time in India studying the sitar. It was then that she had told me about the Ravi Shankar album in the café's collection. The madam had recalled my personal musical history and tailored the café's atmosphere to me, an inclusive and welcoming gesture.

SOUND AS AFFECT MANAGEMENT

The affective sensations I experienced as a patron at that moment in Mignon gesture to the crucial role of sensory management by the madam or master. Music café masters are known to be authorities in their field by "reading" the listening atmosphere and selecting music that will appeal to their patrons. At other times, they simply pick music randomly, as some owners told me. The master at Violon shares his knowledge freely and openly with his patrons.[26] One regular at Violon told me that he always tried to sit in the lower level, directly facing the speakers. He told me the master recommended this seat since it offers the best acoustic experience. Another regular patron told me she liked the café because it was "easy to enter" (*hairiyasui*). She attributed this to the friendliness of the master and madam (his wife operates the turntable during midday hours), after admitting she had been nervous when she first began visiting the café.

Sound also plays a key role in the affect management of patrons. Many café owners told me that they know the tastes of their regular patrons. One told me, while pointing out a patron sitting at the back of the café, "Well, if I see him here, I know that he likes Mozart, so I might play that." Another café madam told me that while she does not always remember the patrons' names, she knows their individual tastes and preferences and will often identify the patron by his or her favorite composer—he's "Bach-san" or he's "Beethoven-san" as she told me. As David Novak writes regarding the jazz café master, "the master's mastery is represented by the ability to 'channel' the correct music for his guest" (2008, 20). Occasionally, however, this impulse to channel and tailor music for individual patrons can conflict with maintaining an appropriate ambience in the café. The madam at a popular café in Tokyo told me that while she has *gendai ongaku* (atonal postwar music) in her collection, she always feels nervous when a patron requests it. Laughing sheepishly, she confessed that it might drive other patrons away. During an interview with the madam of a café that has been in business since 1955, we discussed the music in her café's collection. I asked her if she had any recordings by Satie (most café collections stop after the Romantic era in the early twentieth century). When she answered yes, I told her that he was one of my favorite composers. After our interview ended, I remained in the café for a while longer to take notes and was pleased to hear the familiar strains of Satie's *Gymnopédies* filling up the space.

At another café, an elderly madam likewise attends to her patrons' tastes with detail and careful attention. The first time I went there, I was taken aback by her attentiveness toward the patrons sitting quietly in the café. She filled my water glass multiple times and was constantly giving me fresh warm hand towels (*oshibori*). At one point there was a brief pause between recordings. Despite what one might expect with "dead air" (anathema to live recordings or broadcasts) permeating the space, the madam came into the main space of the café and asked us directly if we had any requests. She spoke to me using the extremely formal and polite form of Japanese speech, noting that "since I'd come from so far," she would like to play a recording of my choosing. An elderly regular patron talked with me about my request, Schubert's *Death and the Maiden*, and we exchanged contact information for a possible future interview. When I first arrived at the café that day, I noticed that a middle-aged man had brought his teenaged son with him. They both sat quietly in separate seats across from each other, both staring intently at the speakers in front of them, barely moving. Later, the madam would tell me excitedly that the man had been a regular patron when he was younger but had not returned in many years. He was eager to bring his son to experience the music café firsthand.

ACOUSMATIC SOUND

In most cafés in Japan, background music serves as acousmatic sound and is projected from hidden speakers. Sound fills the atmosphere, yet its presence is rarely commented on or noticed. Its origins obscured, the music retains a sense of "sourcelessness" (Kassabian 2013, 9–10). In contrast to this structure, music in classical music cafés is radically deacousmatized through the acoustic and material design of the space itself. Café owners are able to structure patrons' experience through this deacousmatized quality of sound.

The spatial design of music cafés is arranged to bombard and surround patrons by sound, as seats are often placed to directly face the massive speakers, thus encouraging certain listening comportments that often resemble those of audiences at live performances. Even though many patrons may be listening to the café music in a distracted fashion as if it were BGM, they are still actively engaging with sound. I recall a young customer who would spend several hours in the café writing in her journal while periodically putting her pen down and staring into space, lightly tapping her fingers on the table in tune with the rhythm. Like many patrons I spoke with, this young woman was drawn to this particular café for its unique and nostalgic ambience. Her listening practices—drifting between distracted and concentrated—resembled those of many other patrons I witnessed, whose auditory practices are directly informed by the deacousmatized presence of the music itself.

Rather than engagement with the music being a type of reduced listening (where the focus is on the "sound itself"), the origins of acousmatic music are foregrounded as fetish objects—the high-end audio equipment, especially the speakers and turntable, the record collection, and the master or madam openly operating the equipment.

At some cafés, listeners face speakers but keep their backs to other patrons. They are often shut off from other patrons even though they are sharing intimate behavior (eating, sleeping, reading, writing) while in close proximity. This is similar to behavior in other cafés in Japan's cities, such as cat cafés, where patrons share intimate yet nonlaborious activities such as sleeping, sitting quietly, and reading, even though they may not be directly interacting with one another. Most regular patrons I spoke with seem to have a preferred seat in the café. This can be an ideal location in terms of the café's acoustics: such seats are usually in the center of the room directly facing the speakers.

The notion of *kodawari* is commonly applied to *kissaten* culture, which refers to "the desired quality of focus and perfection-seeking." This is the primary goal for café masters and madams (and craftspeople more generally), and they are seen as the embodiment of the "desire for the ultimate coffee experience" (White 2012, 66). Interestingly, a regular classical music café patron referred to this concept in relation to the patrons' desire for ideal sound. As he told me, the patrons' preference for seats that directly face the speakers (considered the acoustically ideal location in the café) can be seen as a type of *kodawari*. Or, as another regular patron told me, his preferred seat is directly in front of the speakers, but also situated in its own tiny private nook, so it is out of the way of other patrons and "won't disturb" others, he explained. In this case, establishing a sense of privacy overrules the focus on an acoustically ideal experience. This patron was initially hesitant to be interviewed because, as he sheepishly told me, "I'm not at all knowledgeable about classical music." Instead, as he revealed to me, he goes to the café regularly (several times per week for the past ten years) because it is a quiet space where he can read alone, while the classical music serves as pleasant background. The master and his wife know him well and chatted with him before he left for the day.

Media culture in contemporary Japan has often been described in terms of "mediated and live," in which recordings (the ultimate form of acousmatic sound) and live music performance are conflated (Novak 2008, 31).[27] Many cafés refer

FIGURE 2.14 Unveiling the fetish. Café madam operating the turntable as seen from the café's listening interior. Similar to café Mignon, the jackets of albums that are currently playing are placed in the window in view of the café patrons. Photo by the author.

to business hours as "concert time," thus emphasizing liveness. Recordings are always played all the way through, and many patrons will face the speakers as if they are seated in a concert hall at a live performance. Lion café features a daily concert program from 3:00 to 7:00 p.m.; patrons there are given a monthly program along with the menu when they are first seated. At many cafés, the album currently playing will be displayed on a board, and patrons are encouraged to peruse the album and liner notes. I often saw regular patrons take albums currently playing back to their seats to examine the artwork and liner notes and to jot down personal notes about the album's details. I recall one occasion in particular during fieldwork when two patrons left the café right when the movement ended. While the timing could have been accidental, it also potentially gestured to analog listening conventions. Such modes of auditory engagement go against fragmented listening practices that Adorno decried as "regressive" and atomistic (that is, only listening to the song's hooks). Leaving precisely when the movement ended also suggested an interpretation of the recording as live performance.

One elderly regular patron, a retired literature professor, avidly takes notes on the recordings as he listens; a kind of music criticism, as he explained it.[28] He told me that he enjoys learning about new recordings from what other patrons request. This tactile and bodily engagement with unknown recordings fits in with the analog atmosphere of the café. Rather than use an application like Shazam to render unknown music instantly accessible, listeners here get up out of their seats, walk over to the request board, and scrutinize the recording as an object itself.

SENSING NOSTALGIA

The 2017 Japanese fictional food-themed drama *Samurai Gourmet* explores the connections between food, taste, nostalgia, and sensory memories. The show follows the everyday life of recently retired Kasumi Takeshi, played by well-known actor Takenaka Naoto, with each episode devoted to a distinct mode of Japanese cuisine. The second episode opens with Kasumi strolling through a neighborhood filled with used bookstores and waxing nostalgic about his college days. His youthful memories are steeped in rich sensorial associations of food and leisure. "Ah yes, that smell!" the inner monologue voice declares excitedly, while Kasumi walks through the bookstore aisles. After he purchases a used book, he decides to read it leisurely at a "pure café" (*junkissa*), the type of café he used to visit as a college student. Walking through the streets, Kasumi stops in front of Segafredo, an Italian-themed café chain, and scoffs, "Not this kind of café." He

FIGURE 2.15 Regular patron inspecting and taking notes on the recording. Photo by the author.

describes the pure café in terms of its atmosphere and design: antique lighting, relaxed and comfortable atmosphere, silver teapot filled with sugar packets at the table, watching the café master prepare the coffee. Notably, he describes the sonic atmosphere of the pure café in terms of its BGM without hesitation: "classical."[29]

Described by *Thrillist* magazine as "food porn," fans revel in the show's languid pace, emphasis on food in terms of eating and preparation, and simple plotlines.[30] Indeed, the focus is on Kasumi's sensory experience of eating, which is always meaningfully nostalgic. When he is served his Vienna Coffee, a classic coffee drink at retro cafés in Japan, the tone shifts and the camera zooms in on the whipped cream artfully floating atop the milky coffee. Kasumi picks up the spoon and slowly tastes a small spoonful of the whipped cream. As if overcome, he closes his eyes, and the inner monologue declares dreamily, "So sweet." The next shot is a close-up of Kasumi's hand slowly stirring the remaining whipped cream and incorporating it into the coffee. He smiles and takes his first sip, cocks his head to the side, gasps quietly, and exhales in pleasure. With a smile on his face, he eagerly turns back to his book. The entire episode (and series more generally) emphasizes the gustatory and sensory pleasures of food and memories.

In the music café, these sensory pleasures are inextricably tied to the affective pleasures of analog sound. While pure cafés are not music cafés, strictly speaking, they both enjoyed a renaissance during the same era (1960s and 1970s) and are seen as examples of Showa retro culture.[31] As Kasumi is walking through the neighborhood looking for a pure café, he becomes worried that they no longer exist—that chain cafés have supplanted them all.

In her study of *chindon-ya*, street musicians hired to advertise businesses, Marié Abe examines how an older musical form can continue to thrive despite the conditions that gave rise to it having been radically transformed (2013, 94). Her analysis of *chindon-ya*'s contemporary resurgence is illustrative for making sense of the continued patronage of classical music cafés in Tokyo: they are each often explained simply in terms of nostalgic appeal. Yet nostalgia does not fully explain the persistence of these cultural institutions. Both examples draw on sound to create a powerful sense of intimacy for the patrons. *Chindon-ya* musicians perform in city streets, and therefore work to transform public space into a zone of intimacy; they "challenge the physical separation of public and private space; streets become a site of social warmth and affection when home is considered a place of isolation" (Abe 2014, 104). Relatedly, music cafés draw on the tactile nature of sound (in this case, analog sound) to create sociality and intimacy among patrons. Despite their dwindling numbers, classical music cafés serve as affectively powerful leisure spaces for city dwellers. While they operate on the fringes of mainstream cultural activities in Tokyo, I argue that music cafés and the atmospheres they generate reveal something about how people engage with sound in the city. In this case, the fringes illustrate certain attitudes toward sound and listening—something not to be shut out and privately consumed, but something to be enveloped in and surrounded by. Despite their small numbers, classical music cafés are part of a larger culture of attuned listening in Tokyo's leisure spots.

INTERLUDE II

New Experiences in a New City for New Women

Ambient Sound for Refined Women

For all the new women
who live in this new era
We chose the name NEWoMan
as a place to encounter
new experiences

As described in these lines on its website, the department store NEWoMan opened in Tokyo in 2016, marketing itself as offering "new experiences in a new city for new women." Curious to see how the BGM would sound given the highly gendered dimension of the store, I decided to explore NEWoMan.

Just inside the main entrance, after passing through the automatic doors, I encounter the sound of water bubbling, with the faint sounds of the rainforest in the background being projected from a small, stainless steel rectangular device that could almost pass for a waste receptacle. I am immediately struck by the fact that there is no BGM being amplified in the department store's lobby, a ubiquitous convention of background music programming in Tokyo. When riding the escalators, I encounter the rainforest sounds again throughout the ride. As I disembark, I notice an explanatory sign placed in front of a fern.

The sign explains that the department store wants to create a new sonic environment to go along with the new retail space, NEWoMan. This new sonic

FIGURE INTERLUDE
II.1 NEWoMan entrance, Shinjuku, Tokyo. Photo by the author.

FIGURE INTERLUDE
II.2 Sign at top of escalators explaining "hypersonic sound." Photo by the author.

FIGURE INTERLUDE II.3 NEWoMan women's restroom with modernist yet nature-filled layout. An elaborate design, with moss, plants, and branches filling the space behind the sinks and mirrors. Photo by the author.

FIGURE INTERLUDE II.4 Greenery display behind the sinks in NEWoMan restroom on the first floor. Photo by the author.

environment includes the sounds of the Borneo rainforest and gibbon calls that resonate throughout the open floors and intermingle with the white noise of customers' voices and salespeople beckoning shoppers passing by. The department store's design, spilling out of the south exit of Shinjuku train station, is described by the company as confronting store visitors with a "unified feeling as if in a 'city' with no borders." I am intrigued by the store's use of nature sounds, often associated with the genre of healing music in Japan. It feels like a radical choice to completely reject foreground popular music in a department store and instead favor the more amorphous and ambient sounds of nature.

I recently interviewed the sound design company that programmed the ambient music in the women's restrooms in NEWoMan, so I am eager to hear the music in person.[1] After walking down the short hallway to the restroom, I notice that there is some confusion as to where to queue up. The sinks are in the center of the space, with the stalls surrounding the entire perimeter of sinks. However, the placement of mirrors creates a curious environment. The design focus is on refinement as opposed to antiseptic utilitarian order, as one might expect from other public restrooms.

The BGM in the NEWoMan restroom reveals the gendered dimensions of sound design programming; the ambient sounds are part of the restroom's atmosphere, which is staged to appeal to the senses of female shoppers. While in the restroom, I try to record the ambient sound, but this proves futile, as the faint music does not record easily. More significantly, I am constantly thwarted by the loud and intermittent sounds of hand dryers that punctuate the background sound.

THREE

"Feeling Uncomfortable Without Sound"

Muzak as Affect Management for Office Workers

> If there are spaces without background music,
> one feels uneasy or out of place.
>
> *USEN manager in Japan*

As we saw in the previous chapter, in interviews with music café patrons, many described the pleasures of being surrounded by sound. Some BGM companies in Japan, recognizing profit potential, are taking note. These companies reposition the allure of being cocooned in analog sound in the music café into a discourse that rests on the necessity of ubiquitous sound. Namely, background music companies are now framing spaces without sound as discomforting. The music in the work environments targeted by BGM companies is not for leisure or aimless listening, as practiced in the music café, but instead is for stimulating productivity. The music café's unhurried ambience and foregrounded music is a stark rejection of chain coffee shops' hyperefficiency model, where background music operates strictly in the background. Yet, in a curious contrast, (non-analog) streaming background music is now being employed to structure productivity and make post-Fordist workers happier by nurturing their emotional attachment to work and their co-workers.

In order to examine more ubiquitous examples of sound as atmospheric technique, or sonic patina, I turn to the obvious example of muzak, a form of mass-mediated music that, in its early days in North America, was believed to stimulate, envelop, and target the worker, and later, the consumer. Similar to the ways that classical music cafés provoke competing modes of listening, muzak listening invites a range of somatic responses. Muzak is typically framed as music *not* to be listened to; yet this non-listening is itself a meaningful form of auditory engagement.

Much of the background music heard in consumer spaces and offices in Japan, referred to as BGM in Japanese, is supplied by USEN, a radio streaming service (similar to Pandora) that translates to "wired" in English. USEN provides businesses and private customers an extensive range of foreground music and original studio compositions, as well as classic instrumental BGM covers of popular songs, old and new.[1] Like air conditioning, muzak generates an all-encompassing sensory atmosphere; however, its spatial presence is necessarily ambiguous and indeterminate. Muzak's free-floating, amorphous, and ubiquitous presence has been described by Gary Gumpert as a "kind of amniotic fluid that surrounds us; and it never startles us, it is never too loud, it is never too silent; it's always there" (cited in Lanza 2004, 5). Perhaps it is this manufactured infantilized calm that incites widespread condemnation of muzak by consumers, workers, and musicians, among others.[2] Additionally, Paul Allen Anderson points out that this quality of muzak as mood manipulator serves as a threat to some listeners "because it flouts the ideals of autonomous music, concentrated listening, and interior experience" (2015, 819). BGM in Japan is simultaneously unnoticed, dismissed, and appreciated, and is now being used to heal and manage white-collar office workers.

Muzak is a powerful example of acousmatic sound, sound we hear without seeing its source. Muzak is not to be intentionally listened to or engaged with; however, as I discovered, it can be hard to ignore and *not* listen to. Its repetitive deployment within highly consumerist spaces, and now work spaces, suggests that its presence is something more than simply background music. Although there are legal restrictions on amplified sound in public spaces in Japan, largely based on sound control regulations in the United States, they remain generally unenforced or ignored (Dolan 2008).[3] Since the emergence of antiwar rave demonstrations in the early 2000s in Japan, political protestors have increasingly used sound and noise tactically (Hayashi and McKnight 2005), most recently in antinuclear musical protest performances (Manabe 2015). Driven by the pleasure principle, such tactical and anarchic uses of sound, noise, and music performance in rave demonstrations have "introduced contingency and openings into the space of protests that reject the geometry as admired by right-wing military organizations" (Hayashi and McKnight 2005, 106). This contrasts markedly with the expansive and complex use of BGM in public space, which is predicated on affective engineering to maintain its seamlessness and is less about a pleasurable engagement with music proper than it is an atmosphere evoked through background sound. The term BGM is itself indiscriminately used in Japan and does not have the pejorative connotations (as low art or kitsch) that the English term muzak still retains. For example, in a vegetarian restaurant and live performance space in Tokyo, I once saw a flier on the wall for a local independent band whose music was being advertised as "BGM inside the café" for that particular month's playlist.

As some scholars have noted (Jones and Schumacher 1992; Vanel 2013), muzak has typically been critiqued as either the epitome of "low taste" or the sonic product of control and manipulation ("cultural totalitarianism"). Yet Herve Vanel reminds us that in fact, since its inception, muzak has been "synonymous with inevitable progress" and visions of techno-utopia, and not merely "bad music" (2013, 49). In the 1940s and 1950s in North America, muzak was used to make bored or fatigued workers more productive, more stimulated, and more interested in their work. Now, in post-bubble Japan, muzak is being used to sonically smooth over the rough edges of Japan's still stagnant economy for stressed and overworked white-collar laborers.

Muzak has long been used in Japan's consumer spaces to create specific moodscapes in bars, restaurants, stores, and shopping arcades; however, only recently has BGM been deployed to heal and affectively manage office workers through a background music-streaming program called Sound Design for Office. Pithily defined as "mental balance music for workers," this program, hereafter SDO, is described in a USEN pamphlet for potential customers as a concept "based on optimally designing environments with sound." According to company marketing, SDO is directed at offices to improve the mental health of their workers and protect the "body from workplace stress." It is currently being promoted as a sonic counterpart to government-mandated stress checks for office workers that took effect in December 2015. SDO is premised on the idea that music alters neuroactivity, and therefore has the power to affect individual psychology and mood.

Like Muzak Corporation hiring specialists in medical fields to provide authoritative backing of its research in the postwar era, USEN employs a range of specialists in a range of fields who support claims for the efficacy of background music to alleviate bodily and mental stress. For example, they argue that nature sounds, recommended to be played early in the workday, trigger a psychological association with simplicity, which in turn slows down our sympathetic nervous system, thus producing feelings of tranquility. While this use of BGM as a sonic tool for stimulating office workers' productivity and happiness is clearly modeled after the North American Stimulus Progression program, it is also linked to Japan's healing, or *iyashi*, boom, which became widespread in the early 1990s after the collapse of Japan's economic bubble, the Kobe earthquake, and the Aum Shinrikyo sarin gas attack in 1995. Healing goods and services such as music, color therapy, aromatherapy, robot interaction, and pet therapy became marketed as necessary tools for coping with the economic and social anxieties that intensified by the mid-1990s. USEN's marketing of the desire for healing through background music resonates strongly in contemporary Japan in the wake of the 2011 Fukushima earthquake, tsunami, and nuclear meltdown and ongoing recessionary conditions.[4]

The social and economic unease that has spread since the economic bubble burst in the recessionary 1990s has been more palpable in the wake of the 2011 Fukushima earthquake and nuclear meltdown. Rising suicide rates are accompanied by media discourses and moral panics over fears of social disconnectedness, relationlessness, and lonely death (Nozawa 2015). Since 2002, with the weakening of social welfare and the health-care industry following neoliberal reforms by former prime minister Koizumi Junichiro, privatized services and goods have expanded to provide various forms of affective management like healing music, aromatherapy, pet therapy (including cat cafés), and now, I argue, BGM for white-collar workers.[5] This recent deployment of background music in the workplace is part of a larger discursive shift in post-3/11 Japan toward curbing overtime and overwork in order to combat the increase in stress and mental health issues and the rising number of suicides and death from overwork (*karōshi*).

As one USEN employee I interviewed put it, BGM is uniquely suited to manage and care for overly stressed office workers, at a time when the negative effects of overwork on office workers are increasingly on the rise. USEN is able to implement ubiquitous music programming in offices—spaces that some might see as an auditory escape from the noisy sounds of the city, including constant BGM emanating from stores—by drawing on the already established presence of mechanized sounds, noise, and background music in Japanese urban public space.

Drawing on ethnographic fieldwork with USEN employees in Tokyo and an analysis of the company's print media literature concerning its SDO program, this chapter explores the relationship between background music and the affective management of worker stress in contemporary Japan. My focus is on the entrepreneurs and conceptual apparatus undergirding SDO's deployment, rather than on the worker's sense of agency in being subject to SDO. I examine the discourses surrounding the use of BGM in offices as a means to affectively manage office workers and their environments in post-bubble Japan. This use of background music in Japan to affect and respond to mood is part of a broader global shift toward the use of sound, here "neo-Muzak," as "personal care products for affect management and mood elevation." Neo-Muzak in this case refers to streaming services like Pandora or Spotify that offer users seamless "algorithmic or curated musical moodscapes and affective atmospheres" in order to elevate or enhance one's mood (Anderson 2015, 811). The primary difference is that workers in North America create individually cultivated mood-enhancing soundtracks using neo-Muzak as part of a private auditory experience, while workers in Japan are subjected to SDO as a collective sonic experience.[6]

I argue that BGM in Japan is emerging as a form of ambient labor control that exploits a particular mode of sonic engagement. As I will describe here, muzak historically has not been meant to be actively listened to; however, it is designed to have certain effects on consumers and workers. Unlike classical music cafés, where sound is employed to encourage deep listening, which ultimately provokes a range of engagements that are both simultaneously distracted and concentrated, BGM for office workers is used to stimulate efficiency through somatic listening. Intriguingly, while workers are not expected to directly listen to the office BGM, many speak of its absence (when the programming ends for the day, for example) as unsettling or uncomfortable. Following an account of USEN's founding and growth, I will examine how SDO is able to draw on and exploit somatic modes of engagement with sound other than listening, including the workers' bodily movements, neuroactivity and biorhythms, and sonic cocooning. In its corporate advertising, USEN promotes its services as providing "mental health care for the working person," but also as tools to get workers into a "work rhythm." What precisely constitutes a "work rhythm" in recessionary Japan, and how might "the power of sound"—to borrow USEN lingo—facilitate the creation of an atmosphere that mitigates the stressed and fractured rhythms of white-collar labor?

POST-BUBBLE PRECARITY

After the economic bubble burst in 1991, Japan experienced a deep recession from which the country is still recovering. Post-bubble Japan ushered in a new era of flexible labor: major labor restructuring including ending lifetime employment and privatizing social services. Roughly 40 percent of the workforce in Japan is now made up of temporary and part-time workers. Another major shift during this time, as Anne Allison points out, was the collapse of the health-care system (*iryō hōkai*) and the subsequent emergence of the care market (2013, 38). The workers I am focusing on in this chapter—those who are subject to SDO in the workplace—are not economically precarious in the ways in which temporary or freelance workers are. However, it is the perceived increase in their stress levels, suicide, and mental health issues that is the impetus for SDO's office programming and recent governmental health check reforms.[7] According to a survey on the state of employees' health by the Ministry of Health, Labor, and Welfare (MHLW), about 60 percent of workers have "intense stress and 30 percent of those who commit suicide are workers."[8]

Economic and social precarity continue to be exacerbated since current prime minister Abe Shinzo took office in 2012 and initiated neoliberal economic reforms known colloquially as "Abenomics." One of the components of Abenomics is a series of labor reforms and deregulations that involve expanding limited regular employment, a putative middle ground between regular employment and irregular precarious work. Not surprisingly, businesses see this expansion as offering more flexibility while working to slow the division of the Japanese workforce into "winners" (those with regular jobs) and "losers" (those with irregular jobs) (North 2014). Critics argue that Abe's policies will ultimately create more job insecurity and further weaken workers' rights, especially those of women and young people (2014). Abe only recently acknowledged this crisis by proposing a mandatory holiday bill that would begin to tackle the effects of overwork and promote "work-life balance" by requiring workers to take their yearly allotted vacation days, which many people still do not take.[9] USEN's use of BGM for managing and caring for stressed office workers who are under increased productivity demands occurs alongside a broader concern for maintaining work-life balance and regulating the culture of overwork by managing the mental health of workers. The economic and social insecurities of everyday life in recessionary Japan have become more pronounced in the wake of the 2011 Fukushima disaster, leading to the experience of social precarity—"being and feeling insecure in life that extends to one's (dis)connectedness from a sense of social community" (Allison 2013, 348–49)—as the ordinary state of life.

FORDISM TO POST-FORDISM: FROM MUZAK TO AUDIO ARCHITECTURE

The relationship between music and labor has a long and complex history. Music has long served as an ordering and regulating force, attuning workers to the rhythms of the workplace. It was believed to prevent bodily and mental fatigue and boredom from the repetitive nature of work. It is these qualities that putatively allowed music to alleviate worker fatigue, enhance the environment, and prevent distractions that were seen as disruptive to physical labor and repetitive office work in the early twentieth century. The origins of Muzak are rooted in the work of General George Owen Squier, who served as chief signal officer of the US Army Signal Corps. In 1922, Squier developed and patented a means to transmit audio into homes over electrical power lines, naming his discovery "wired wireless." As a response to the privatization of the telephone industry, wired wireless was to be an American public utility available free to all (Sumrell and Varnelis 2007, 107). In 1934, Wired Radio Inc. became the Muzak Corporation and began wiring music directly into people's homes in Cleveland (2007).[10] In the late 1930s, Muzak was first deployed in commercial spaces including restaurants, hotels, bars, and nightclubs (2007, 108–9). With speakers embedded in plants, background music was dubbed "potted palm music" for its acousmatic nature; it was employed in particular in leisure contexts (2007).

Muzak later began to be incorporated into work and factory environments in Britain and the United States to provide aural stimulus for weary factory and office workers during World War II. The supposed efficacy of using music to improve productivity was based on scientific research and testing sponsored by the military that concluded the physiological effects of music on the body: "By affecting the body physiologically, background music could keep workers' nervous systems calm, thereby giving them greater emotional stability during the difficult days of the war" (Sumrell and Varnelis 2007, 112).[11] Drawing on studies in the field of industrial psychology, it was believed that functional music alleviated boredom and increased concentration, thereby making workers more efficient while on the job. Such research linking sound to worker productivity are claims that we see returning, albeit with some important differences, in the recent discourse on BGM for office workers in Japan.

In order to maintain productivity, it was crucial that Muzak be unobtrusive yet familiar; it had to take on an aural invisibility to keep workers from actively listening to the music. If specific songs were requested by workers, they were quickly removed from the library. "Like the Fordist worker," Robert Sumrell and Kazys Varnelis note, "Muzak that drew attention to itself was deemed unsuccessful and dismissed" (2007, 117). By the 1950s, a program known as Stimulus Progression emerged in the United States as a way to keep office workers focused and interested in their work. Based on the human heartbeat (seventy-two beats per minute), Stimulus Progression presented a range of dynamic levels over the course of the program in fifteen-minute segments, taking into account musical tempo and instrumentation. Muzak ranked music in terms of its stimulus value, which determined when it would be played during the workday. The program was designed to follow and regulate the workers' natural levels of attention and arousal throughout the workday. Thus, periods of higher stimulation and mental focus (morning) demanded music with a higher stimulus value than that played during periods of mental sluggishness (lunchtime) (Anderson 2015, 822). Significantly, Stimulus Progression also built in fifteen-minute periods of silence between musical segments in order to keep workers focused and not overwhelmed by constant music with no breaks.

Sumrell and Varnelis argue that the transition from postwar era stimulus progression to audio architecture of the 1980s is indicative of the shift from Fordism to post-Fordism. Muzak no longer serves to discipline the office or factory worker or directly manipulate the consumer's attention, but instead is used to construct audio environments in retail (and, in the case of Japan, work) settings through the technique of atmospherics and the creation of moods. Simply put, atmospherics can be defined as a technique that creates a "distinct ambient audio environment for a particular retail environment" (Sumrell and Varnelis 2007, 125). During the postwar Fordist era of Muzak roughly from the 1940s onward, those behind the music programming were referred to primarily as engineers and programmers, while the musicians who recorded for the company were known as the artistic labor. By the 1980s, Muzak engineers had become "audio architects," a discursive shift that emphasized the aesthetic dimension of background music, a form that is not purely scientific (2007, 124).[12] Another significant shift with Muzak since its Stimulus Progression days was the adoption of foreground music first employed in 1968 by Yesco, a Seattle-based competitor company to Muzak Corporation. Foreground music, referring to original music by original artists, soon became the primary form of programmed music to circulate in retail establishments

(Sterne 1997). Hearing original pop music (by the original artist) in your local Starbucks is a perfect example of the ubiquity of foreground music. Muzak no longer serves to directly control the consumer, but instead to amplify a store's brand identity and create a specific atmosphere.[13]

The idea of silence as conducive to intellectual or creative labor has been increasingly abandoned under flexible post-Fordist labor regimes, as people more frequently work in public communal spaces like cafés.[14] The app Coffitivity, for example, promises on its website to "recreate the ambient sounds of a café to boost your creativity and help you work better."[15] Evoking Muzak Corporation's dependence on scientific (or pseudoscientific) research to support its claims of promoting efficacy, Coffitivity informs us that its "coffee shop sounds are supported by science and backed by a community of people using the app every day." The free version of the app offers a choice of three gradients of ambient noise, all rich examples of sonic patina: Morning Murmur ("a gentle hum gets the day started"), Lunchtime Lounge ("bustling chatter of the lunchtime rush"), and University Undertones ("the scholarly sounds of a campus café"). Users of Coffitivity cite a range of things that make the app effective: it helps them concentrate while working; it provides a comforting aural presence when they are working alone and feeling lonely; it generates ambient noise while they are working in a silent space; and it can mask the sounds of loud neighbors. Interestingly, some users report that the app helps calm them when they are stressed. It is this perceived quality of background noise and atmospheric music that is being mobilized in new ways in recessionary Japan under the guise of caring for workers.

BGM AND SENSORY MANAGEMENT IN JAPAN

One of my informants, a sound artist, likened the presence of BGM in stores and restaurants to that of incense: it provides an aural tint, but no one actively senses it. If BGM were to disappear, however, patrons would notice the absence of this ubiquitous aural backdrop. Many I spoke with echoed this statement; even if we find it irritating, we are somehow dependent on BGM's fluid and diffuse qualities. To understand the ubiquitous and highly engineered presence of BGM throughout Japan's urban centers, as well as USEN's broad market penetration, it is important to first examine the ways that atmospheres have been created in Japan, and in particular, the manipulation of sound in the regulation of such affective atmospheres.

The most obvious reference point is the long-standing national radio exercise program (*rajio taisō*) that began in 1928. Such health-oriented programs were already popular in the United States; inspired by a health/life insurance executive's trip to the United States, the program was created in Japan as a way to tackle national health issues such as tuberculosis and other infectious diseases (Fukue 2009). The executive received support from the physical education chief at the Ministry of Education, the Life Insurance Association, and the Japan Broadcasting Association to broadcast the program nationwide (Faison 2007, 181). Factory workers, office workers, and schoolchildren performed radio calisthenics in unison, being guided by a disembodied voice coming from the radio (Yasar 2009). Banned during the United States occupation in 1946 due to its militaristic connotations, the program reemerged in 1951 on NHK, Japan's national broadcasting network (Fukue 2009). Promoted as simple exercises to boost and stimulate blood circulation and flexibility, radio calisthenics are meant to be performed in the morning, before work and school. Twenty percent of Japan's population continues to perform the exercises (2009).

There are intriguing parallels between these exercises facilitated by disembodied sound (voice; simple melodies on the piano) and the basic premise of the SDO program, where acousmatic background music is engineered to conform to workers' bodily movements and neuroactivity but simultaneously manipulate. Notably, SDO features a twenty-four-hour radio calisthenics channel for workers that is recommended by USEN to be played in the morning, afternoon, or evening, during times when you want to relax and "loosen up your body." This exercise program is followed by the "refreshing" sounds of "twittering birds."

BGM sponsored by USEN started in 1961 during Japan's rapid postwar economic expansion. According to my informants at USEN, early company history is not officially recorded anywhere; however, several employees I spoke with recounted the beginnings of BGM in postwar Japan.[16] The impetus for USEN was born of a frustration with background music listening in Osaka's public leisure spaces. As the story goes, USEN's founder would often spend time drinking at hostess bars, where the women working there would sit with and talk to the patrons as they drank.[17] Music was always playing on a record player, and when the first side of the record ended, the hostess would have to get up, interrupt the conversation, and change the record to the other side before resuming her banter. During this period, conversation was impossible—a disruptive and inevitable puncture during these leisurely social interactions, according to USEN's founder. He had thought it would be ideal if there were a way to play music automatically with no interruptions or gaps between record sides or tracks.[18] This corporate historical narrative is intriguing for the ways in which it invokes the importance of sound not only in creating an atmosphere of healing, but for facilitating communication between hostess and patron. As I will show, it is this quality of sound that companies using SDO claim is one of its primary benefits: strengthening communication between workers while simultaneously masking unwanted sounds of labor from their co-workers.

During the first years of USEN's operation, the company featured only two channels. By 1977, it had increased to twelve channels, but did not have much genre variety, according to a USEN employee I interviewed. At this time, the programs were primarily featured in nightlife leisure spaces such as Japanese-style pubs (*izakaya*) and restaurants; USEN had not yet expanded its services to supermarkets, perhaps the most classic venue for background music.[19] By 1987, USEN experienced phenomenal growth and expanded to 440 channels. Like Muzak Corporation's rebranding after its 2011 takeover by Mood Media, USEN has rebranded BGM from kitschy low art to a complex, intensely curated stream of channels, featuring painstakingly minute taste selections for bars, restaurants, shops, medical offices, banks, homes, and now, business offices.[20] In all of these cases, BGM serves to create an atmosphere conducive to lounging, shopping, or working, or simply one that helps a store further consolidate its image.

MUSIC WHILE YOU EAT

In an insurance office pictured in a USEN promotional brochure, the workers are eating lunch, perhaps to faint music evocative of Paris. The office resembles the design of many contemporary Japanese offices—the desks are open-air and shared by several employees (no cubicles or private work enclosures), the walls and desks are blindingly white, the lighting is fluorescent, and the carpet is a neutral grey-tan color. The interior is cool and stark and is matched by the monochromatic attire of many of the employees. There are speakers mounted in corners of the main work space as well as in the break room. It is now between 12:00 and 1:00 p.m., so the BGM should be cycling through the "refreshment" segment of the recommended workday set: "light and refreshing" music that, USEN claims, "maintains your mind-body health through a change of pace."

USEN's channel H-17, curiously titled "Colorfully Fruitful BGM: Our Company's Cafeteria," reverberates throughout the insurance office.[21] According to the channel description, "The light and organic sounds of things like accordions and violins brighten the company lunch canteen and cafeterias to almost make it feel like you are enjoying a wonderful lunchtime on the streets somewhere in France."[22] H-17, we are told, "enhances refreshment and enjoyment during lunchtime, adding color, and a soothing ambiance to your meal." As I listen to a song from this channel, I note that the warm and resonant sounds of the accordion and bass clarinet, while gesturing vaguely to an imagined idea of France, seem to counteract the cool, antiseptic ambience of the office. The juxtaposition creates a leisurely feeling for office workers as they prepare for the hectic pace of the late-afternoon workday.

USEN introduced SDO in 2013 during the company's fifth decade. During a recent interview with USEN employees about SDO, they were quick to emphasize that this program is a wholly new mode of using BGM and has been "selling like hotcakes" (*bakaure*) since it first emerged. Prior to SDO, BGM was limited to retail venues. As with the postwar era Stimulus Progression program and the more recent Coffitivity app, SDO's efficacy is backed by authoritative figures such as psychiatrists, medical doctors, occupational health specialists, and composers. Work-life balance consultants and design experts also vouch for USEN; such specialists supervise many individual music channels. Intriguingly, USEN employees themselves serve as test subjects for SDO. I was told that USEN tested SDO programming on its own employees before rolling out the program. And yet the experiment is endless, as SDO enters the "real-life" labs of white-collar offices and the data ultimately cycles back into future USEN programming designs. USEN posts the results of such experiments in its brochures for new clients. The manager I spoke with conceded in an interview that the workers are probably not directly listening to the music, but nevertheless, it still has "effects." USEN confirms these effects through in-house testing, which examines the reduction of stress using music and measured physiological responses such as body temperature, sweat, and saliva.

I was told that the exact number of companies that subscribe to SDO is a company secret. Yet what we do know is that the companies vary widely in size, ranging from fewer than ten employees to eight hundred; and type, ranging from the service industry (advertising, travel), law offices, and education (high school) to technology and communications (IT, printer technologies). Indeed, IT companies make up a large portion of USEN's clientele base.[23] USEN employees said that increasingly larger companies are subscribing to SDO, and that for them, music is just one component of managing the work environment. For instance, in addition to background music or sound, these larger companies will also employ aromatherapy, massage chairs, and green spaces for workers. At the time of research, companies paid a monthly subscription fee of 5,000 yen, which gives them full access to the entire SDO set of seventy-six basic channels that they can program to their own liking, though USEN recommends that specific channels be played at certain times of the workday. In addition to the music streaming service, companies rent the speaker system directly from USEN, unless their offices are already equipped with speakers embedded in the ceilings.

When SDO first emerged, USEN promoted and advertised the program to companies as general awareness of the effects of stress and mental health issues for workers began to increase. However, under the specter of impending mandatory stress checks, the program began to receive more advertising coverage, and thus, more clients. Potential customers will meet with USEN's program coordinator to learn how SDO can be applied to their work environments. For instance, I was told that in addition to basic information such as company size, the coordinator will try to ascertain a sense of the employees' work schedule (when they take lunch, and so on) and ask about the company's gender makeup. For offices with more female workers than male, USEN will recommend "stylish" (*oshyare na*) music such as café music, lounge, bossa nova, or jazz. This is done because USEN believes, I was told, that women are more "sensitive" (*binkan*) and will be more receptive to such genres.

BGM AS AFFECTIVE MANAGEMENT

USEN's programming prevents listeners from taking pleasure in SDO as simply entertainment; instead, the BGM is a structural component of the work environment designed to manage workers' moods and productivity. SDO promises to restore "work-life balance" by caring for the working people's bodily conditions and their "weary spirit" (*tsukareta kokoro*), healing them while strengthening their concentration at work. It is not surprising that USEN would harness BGM as a means to restore work-life balance at this particular moment. As previously mentioned, panics over "death from overwork" (*karōshi*) have existed in Japan since at least the bubble economy era of the 1980s. However, in the wake of the 2011 Fukushima earthquake and nuclear meltdown, rising suicides rates, increasing social and economic precarity, a falling birth rate, and a growing elderly population, the instability of white-collar workers has received greater scrutiny by the Japanese government.[24] The government is currently attempting to curb the problem of overwork (and thus restore work-life balance) and overtime work (*zangyō*), both of which are viewed as "threats" to increasing the birth rate.

With the emergence of mandatory stress checks for company employees based on recent revisions to Japan's Industrial Safety and Health Act, USEN is now directly marketing SDO as a sonic means of helping companies meet these labor reforms. In a confidential brochure for SDO clients given to me by USEN, the first few pages provide data concerning the increase in mental health issues and suicide by workers from 2007 to 2014. The brochure then explains the new mandatory stress checks, effective as of December 2015. Following this, SDO is introduced to potential customers under the subheading "proposals to improve work environments through music." The background and impetus for SDO is explained as a desire to explore the possibilities of sound at this "turning point" in which psychological disorders are on the rise in today's "stressful society." In addition to invoking widely discussed social problems like suicide and death from overwork, USEN references the negative outcomes of workers' mental stress in its monthly promotional magazine by invoking the experience of "depression" or other seemingly less serious mental health maladies, such as lost sleep and "May sickness," the annual ennui that occurs among office workers and students every spring.[25]

Sensory regulation through BGM, according to USEN's advertising discourse, creates the proper environment through which to make work more manageable and less stressful. SDO also promises to improve communication among workers, which would suggest an existing lack of connectedness between employees (or at least the feeling of this). As a USEN manager's assistant lamented in a personal interview with me, it is becoming more commonplace for workers to work at their computers without any direct physical contact with their co-workers. Such lack of direct communication is especially stressful among new employees who are recent college graduates. It is difficult for many of them to ask questions or talk to their co-workers, she explained, but if music were present, it would help them in talking with others in the workplace. The logic, according to USEN, is that the presence of background music will fill up empty spaces in the environment, therefore creating a softer atmosphere. Many of the companies using SDO describe how the program facilitated better communicative practices among co-workers.

Management techniques are exploiting the possibilities of sound at a time when "relations between physically present colleagues have become subject to instrumental commercialization" (Gregg 2010, 267). Affective management instrumentalizes sound, spatial layout, and the textures of a workday's flow. The notion of an affective atmosphere is inherently ambiguous and indeterminate; it exists "between presence and absence, between subject and object/subject and between the definite and indefinite" (Anderson 2009, 77). Atmospheres suggest a specific spatial orientation for the ways in which they fill up, surround, and envelop people, objects, and space: "diffusion within a sphere" (2009, 80). In this case, the affective atmospheres created by SDO programming draw on the spatial orientation of atmospheres as "interlinked with forms of enclosure and particular forms of circulation—enveloping, surrounding, and radiating" (2009, 80).

In dramatic contrast to the standard North American office layout of partitioned cubicles with workers physically and sonically demarcated from their co-workers, Japanese offices are traditionally open-air with all employees, from the lowest to the highest level, working together at individual desks in the same shared space.[26] This spatial openness extends to the workers' collective engagement with sound. Unlike North American workers, many of whom use Pandora or Spotify playlists for personal mood elevation while at work, workers in Japan are all subjected to the BGM that permeates their shared office space. Theirs is not a private auditory experience. The open office layout in Japan can allow for greater mobility and communication among workers, but also greater circulation of unwanted background sounds (co-workers talking, people typing, machinery).[27]

Channel C-71 is named "Office Speech Masking," and is promoted for use in reception areas or lobbies—in other words, in places where workers want to avoid having others hear the details of their conversations. This channel features nature sounds of the "murmur of a stream" as the sonic base over which the audio of lively conversations is overlaid, thus providing a masking quality.[28] C-71 can also be used in spaces where BGM (that is, instrumental or vocal music) would not be appropriate, such as conference rooms. However, as the description for C-71 notes, depending on the circumstances, it is also possible to mix "easy-listening" music with speech masking. By contrast, USEN's other masking channel, H-09, offers a higher level of masking and is based on a different logic for masking the sounds of "offices overflowing with noise" by focusing on instrumental music rather than the sounds of nature. Assuring listeners that the music will alleviate stress for the workers, H-09 plays "beautiful melodies" on the strings and piano that impart an "elegant and gentle atmosphere."

USEN conducts extensive market research on its clients' experiences with SDO and posts the (largely favorable) reports on its official website. One such client, a law firm, explained that it initially expected to use SDO only for sonic masking purposes, given the importance of protecting its clients' privacy. However, as the law firm managers explain on USEN's website, given the range of effects that SDO can have on one's mental health, they realized that subscribing to SDO was the right choice. This use of sound as "techno-cocooning, using technology for creating sensory privacy" (Bijsterveld 2010, 192) is a notable component of how managers and companies are using music and other sounds to create space between workers.

Many of USEN's clients speak of SDO's impact on the work environment as "immediate," and as having created a "softer" (*yawarakai*) work environment. Companies claim that their atmosphere has improved, and that the office itself feels lighter and brighter. Yet they rarely identify specific genres or artists as helping create this changed environment. More important, USEN managers and SDO subscribers frequently speak of an increase in work "efficiency," or *merihari ga tsukimashita*. The phrase *merihari ga tsukimashita* implies a sense of rhythm regarding the smooth functioning of the business. To maximize *merihari ga tsukimashita*, companies seek to create a rhythmic space where people can work effectively and efficiently. Writing in USEN's promotional magazine, the manager of a major real estate company, Mitsui Home, claims that since he has been subscribing to SDO, workflow has become more efficient. He explains, "For instance, in the morning you have the loud clattering of keyboards ringing into the quiet space, but now that there's music playing, we're able to work more smoothly." Here, sound is believed to increase productivity and worker efficiency through covering up or aestheticizing the unwanted sounds of labor.

Today in Japan, sound is positioned as a necessary part of the workplace's spatial and affective design. For many clients, it has become an indispensable feature of the work environment, such that some workers claim to be unsettled by the mere thought of a workplace without the omnipresence of SDO throughout the workday. In USEN's promotional magazine, an employee whose company uses SDO noted the dramatic change in ambience, stating that everyone began to communicate with each other more easily. She revealed, "Now if I just think about the office without SDO, I feel unsettled [*iwakan*]." Similarly, the USEN manager I interviewed told me about a time when he worked long after the SDO programming had fully ended and there was no BGM. He was surprised by how noisy the office became (with the sounds of typing, and so on) and how unsettling (*iwakan*) the feeling was without ubiquitous music in the background. The section manager for a major telecommunications and media company (over eight hundred employees) explained that prior to subscribing to SDO, employees had difficulty going to public spaces meant for socializing, such as cafeterias. Without sound, these spaces seemed "lonely or desolate" (*sabishii*) and thus created a foreboding atmosphere. By contrast, with SDO, communal spaces such as cafeterias became more welcoming. In them, workers could communicate and socialize with one another without having to lower their voices. I interviewed an employee at a life insurance company that was one of the first SDO customers. She confirmed the feelings of being unsettled when the music ends for the workday: "If I go in and work on a weekend or on a holiday, there's no music playing in the background and my first thought is that it's so quiet!" She added that this absence of BGM made it difficult to concentrate and be productive. To her, the BGM-less space evoked the feeling of being alone (*kodoku na kanji*).

Spaces without sound are being engineered and framed by USEN as desolate and unsettling. It is worth noting that the Japanese term for feeling uncomfortable or unsettled, *iwakan*, also evokes a sense of incongruity or incompatibility. In this context, BGM (and its power as diffuse, affective sonic material) is being exploited to neutralize and normalize work environments in order to transform workers into happier, more stimulated, more relaxed, and, most important, more productive workers. The act of listening to the SDO program is somewhat downplayed; instead, sound is envisioned as *filling up* the space and suffusing the entire work environment, fluidly adapting to the distinct affective demands of each space (lobby, work space, cafeteria, and so on). To create this effect, USEN's speaker systems are designed so that the volume between songs is continuous and steady, with no gaps or jarring volume changes between tracks. This sonic continuity is crucial to sustaining the worker's engagement with sound throughout the workday.

USEN selects and manages its genre channels based on information provided by companies about their work and consumer environments. The USEN manager told me laughingly, "Sometimes if I go out drinking at night and I find myself annoyed by the [USEN-sponsored] BGM that is playing, as head of my section, I'll call up the [lower] staff members and yell at them for choosing the wrong BGM!" During our interview, this manager and his assistant spoke precisely and eloquently about what music genres fit with what environments. They both told me that they are always conscious of how BGM is used by various businesses. "We're all music fanatics (*otaku*) here!" they told me. During the interview, the manager punctuated our discussion of the specific channels by playing them for me. We met in a private room that would typically be used for potential USEN clients; thus, the space was outfitted with a speaker system to preview channels for customers. "Put on C-22," he told his assistant, as we discussed the healing music channel. Instead of merely explaining that this channel is simply the sounds of waves crashing, he wanted me to hear it for myself. "We have a lot of these type of channels," he told me. Not music at all, only different categories of nature sounds, all for "creating atmosphere [*funiki*]."

Similar to the earlier North American–based Stimulus Progression program, SDO offers companies a structured music program that paces workers throughout the day. Managers at USEN are aware of this historical connection, as was revealed to me during an interview. Still, there are important distinctions between Stimulus Progression and SDO and their corresponding labor rhythms. While Stimulus Progression was structured to follow the Fordist workers' natural ebbs and flows of energy throughout the workday, its primary impetus according to Muzak corporate literature was to "combat monotony and offset boredom" (cited in Lanza 2004, 49). SDO, on the other hand, engineers BGM to help employees cope with the precarious conditions of post-economic bubble Japan and its ensuing care deficit. The collapse of the economic bubble saw the dissolution of Japanese management style and a shift toward post-Fordist practices of flexible labor resulting in widespread social and economic precarity. So-called Japanese management style refers to "lifetime employment, wages based on seniority rather than merit, and all-company unions," but Anne Allison also points out how this management style "indexes those business strategies intended to foster sticky socio-affective ties between worker and workplace" (2013, 25). SDO employs continuous sound to reinvigorate these weakened affective ties between worker and workplace. For example, channel K-17, part of the awareness category, is titled "Morale Building ~ Theme from Rocky," and consists solely of the theme from the film *Rocky* played on a continuous loop. The efficacy of this channel was pointed out to me during a recent interview with a USEN employee. It is also highlighted in USEN's promotional literature by companies that program K-17 at the end of the workday as a sonic mode of team-building and solidifying workers' pride.[29]

Organized to follow the natural ebbs and flows of attention and concentration throughout the workday, SDO targets four distinct categories of affective states: concentration, relaxation, refreshment, and awareness. Each category offers a variety of individual channels for users to select. BGM is broadly defined in this context—programs feature the work of established classical composers alongside original music (both classic and popular) and newly written compositions by USEN's in-house composers, as well as ambient sounds of nature. Similar to the attention toward choosing Stimulus Progression music that would not distract workers lest they become too attached to the music, SDO programming features only instrumental music, since vocals would be "too distracting," I was told. If the music has vocals, then workers will start to listen and will not be able to concentrate on their work; they will become too aware of the music's presence. The awareness category is an important exception here, as it is the only category to feature programming with voices. In this case, the voices are meant to be consciously attended to and processed by workers, not registered as background, as with the other SDO instrumental channels.

Typically played during the late morning, the concentration category (*shūchū kōjō*) features channels designed to improve workers' concentration and boost their productivity. The music featured on these channels is a mix of gentle classical music, focusing on composers including Bach and Mozart, and office noise (*zatsuon*) masking. For example, channel C-65, titled "Office BGM," is described as "Helping to create a soft and pleasing ambiance for working people through the thoughtful presentation of songs selected for each hourly block to match the different biorhythms that people have. This will create an environment that is easier for people to work in." While there is no explicit mention of matching the music's tempo to workers' heart rates, this feature was highlighted in Stimulus Progression, where the workers' bodily movements become supported by and regulated to the sonic environment. The description notes that this channel is designed to "match the moods and rhythms of working people by playing music with tempo."

As I listen to an audio sample from channel C-65, I am surprised by what I hear. It is instrumental with soft, nonthreatening guitars and piano repeating a simple melody three times before it resolves. There is the faint sound of wind chimes underlying the guitars and piano. The musical textures and melodies strike me as more fitting for the atmosphere of a non-place such as a waiting room or passageway, and not as music explicitly designed to help workers concentrate as they sit at their desks. While classical music is also well represented in the concentration category, the inclusion of this soft instrumental channel in the segment of channel C-65 I sampled suggests that USEN promotes workers' concentration partially through creating calm and pleasing sounds. Indeed, the purpose of the music is not necessarily to stimulate their minds, but to wrap them in a cocoon of sound that is in step with their senses, neuroactivity, and bodily movements. The workers must cultivate a different somatic mode of engaging with sound other than concentrated listening. Their primary mode of engagement with this specific channel is through acoustic cocooning; they are enveloped in sound that matches their work habits and movements. The description of this sonic blanket in company literature is based on the therapeutic effects of sound, claims that are backed up by the various specialists who vouch for the efficacy of SDO on workers.

The concentration category is directed by a medical professional who specializes in the therapeutic effects of music.[30] He describes the hindrance of stress, whether social or psychological, on the worker's ability to work and concentrate. He wants to keep stress at a level where workers are not forced to rely on their sympathetic nervous systems—their "fight-or-flight" instincts—to kick in, which will negatively affect digestion and cause blood flow disturbance. Ultimately, he argues, it is necessary to play music that stimulates the parasympathetic nervous system ("rest and digest"), which will save energy by lowering blood pressure, slowing one's heart rate, and starting the process of digestion. Listening to this music will remove mental and physical fatigue and allow workers to concentrate and be more productive. This is sonically achieved by alternating high-frequency sounds with simple and bright short melodies. By playing to the workers' unconscious reactions to the music within the concentration category, SDO programming invokes Brian Massumi's correlation of affect with intensity as "embodied in purely autonomic reactions most directly manifested in the skin—at the surface of the body, at its interface with things" (2002, 25). The compositions featured on this channel are entirely USEN originals—described as "contemporary music that draws on the 'Mozart effect'"—composed and performed by the company's in-house pianist and composer, Nakamura Yuriko.

The relaxation category (*rirakkusu*) offers channels that serve to "alleviate stress for the workers." The category description acknowledges that it is "easy to feel stress working with many other workers in a cramped space."[31] Although workers are collectively exposed to sound within a confined space, the music aims to open up the space between workers and fill up the blank spaces with sound. The relaxation segment is typically featured in the morning hours of the workday, roughly from 9:00 to 11:00 a.m., just before the concentration segment of the workday. Individual music channels include healing music, relaxation guitar, relaxation piano, meditation music, mental care, gentle classical music for working mothers, landscape music (whale sounds), and five varieties of healing sounds of nature to choose from (ocean, forest, clear stream, birds, and insects). The primary goal of the relaxation category is to alleviate mental and physical stress for the workers in preparation for the more intense period of the workday.

Ambient or new age music is favored, and tempos are typically slow to medium, if not arrhythmic nature sounds.[32] The music generates gentle and protective atmospheres for the workers, as if to momentarily transport them—whether to the forest, ocean, Parisian streets, salon, or somewhere more abstract and internal—before the hard work of concentration begins.[33]

The refreshment category (*rifuresshu*) promises to maintain your mind-body health through a change of musical pace. USEN recommends that it be played from early to late afternoon. This is a time in the workday when conversations are active, according to the USEN employee I spoke with, so the music's tempo and pace should reflect that and allow for easy communication. Many of the channels in this category highlight the importance of "workplace communication" and favor music that creates a "light and refreshing atmosphere" for the office. Channel A-10 offers "relaxing music that refreshes the mind and body." The music in the refreshment category includes classic "easy-listening" music (Henry Mancini, 101 Strings, and so on),[34] resort music, several jazz channels (each featuring specific eras such as hard bop), acoustic sounds that are described as being "gentle on the heart," and gentle classical.

The least obviously musical of the four categories, and the only one to feature vocals, is the awareness category (*kizuki*). Promoted as a "useful sound seal that supports mental health measures," this category is made up of various sonic disciplinary announcements and pre-recorded jingles, ranging from weather reports to the happy birthday song to radio calisthenics (*rajio taisō*), meant to be broadcast throughout the office. The category also includes general reminders to the workers, such as not to work overtime hours on designated days. Channel J-35, to be played on a loop toward the waning hours of the workday, features the classic song "Farewell Waltz" ("Auld Lang Syne"), which has long been used by stores in Japan to announce that they are closing, and is thus widely known to sonically evoke "closing time." Tapping into the power of sound to compel listeners, the description of J-35 explains that when most people hear this song, they will automatically think, "I have to go home." Also part of the awareness category is a similar channel signaling the end of the workday that features Antonin Dvorak's *New World Symphony*. The piece's gentle nostalgia, the channel description claims, will produce feelings of "work satisfaction" and "adequacy."

Preceding this channel is the most overt and explicit anti-overwork channel in line with recent discourses in Japan about the perils of overwork and the problem of maintaining proper work-life balance. Named the "No *Zangyō* Day" (no overtime) channel, it is broadcast twenty minutes prior to the end of the workday, just before "Auld Lang Syne" and/or *New World Symphony* are played. It serves as a sonic disciplinary tool to remind workers to leave when work ends, so that they can "be with their family," as an employee told me. The program lasts for five minutes and repeats on a loop.

Company literature notes that the notion of "no overtime" days has recently increased throughout numerous companies, but in reality, most people still work (unpaid) obligatory overtime hours, with many refusing to leave before their superiors. The "No *Zangyō* Day" program firmly instructs the workers in a "gentle female voice" in order to create an environment that "makes it easy to go home." The loop begins with backing synthesizers bursting in with a cloying melody that punctuates and foreshadows the announcer's voice: "Great job today! Today is a no *zangyō* day. Let's all go home soon!" USEN hired well-known voice actress Hidaka Noriko as the voice of the "No *Zangyō* Day" channel. It is not surprising that a female voice is used; the voice, diction, and affective quality are identical to the acousmatic voice announcements in department stores to proclaim their closing. The tone and sentiment—encouraging yet gently forceful, if patronizing—is familiar to Japanese listeners. Pre-recorded announcements occur regularly throughout Japan's sonic landscape, especially in train stations and department stores. The accompanying description for the "No *Zangyō* Day" channel, written by an occupational health physician, describes the perils of overwork, noting that as overtime has become more commonplace, workers are sleeping less, often no more than five hours. The physician explains that overtime continues to occur partly due to cultural reasons, as "if a worker sees that their superior is still working, they will think that it will look bad if they go home before him." Therefore, with the "no *zangyō* day" announcement, everyone will (presumably) go home together.

CONCLUSION: "PEOPLE ARE ENERGIZED, BUSINESS IS ENERGIZED"

USEN envisions workplaces, leisure spaces, medical offices, and households fully immersed in continuous sound, whether background music, sounds from nature, or announcements. According to the company, sound is vital to smooth relations and affective contact (*fureai*) between workers and their superiors. As it promotes in its advertising (and as confirmed during my interviews with employees), USEN positions spaces without sound as physically detrimental—one's stress will increase without appropriate background noise, and the resulting tension will affect and weaken one's concentration. One's productivity may also be negatively affected in a silent workplace, the company literature states. Work or home spaces without intentional background music or sound are believed to be unnatural, unsettling (*iwakan*), and desolate (*sabishii*), and, by extension, unproductive. Throughout USEN's messaging, worker stress is framed as a threat to the smooth functioning of the company. As the cover story of the 2015 issue of USEN's promotional magazine proclaims, for instance, "When workers are energized, business is energized." Through SDO's continuous, enveloping presence, USEN is engineering sound and background music to compel workers to cultivate somatic modes of engaging with the sonic that are different from direct listening. Workers are encouraged to refine and nurture their emotional attachment to work and their co-workers.

USEN's adoption of BGM to heal frazzled and overworked employees can be seen as an extension of muzak's roots in notions of progress, techno-utopianism, and sensory engineering. Affect is particularly heightened and targeted within everyday city life as "part of a reflexive loop which allows more and more sophisticated interventions in various registers of urban life" (Thrift 2004, 58). Nigel Thrift points out that "these knowledges are not only being deployed knowingly, they are also being deployed politically to political ends; what might have been painted as an aesthetic is increasingly instrumental" (2004, 58). Here, we are reminded of the neuropolitics of SDO programming "beautiful music" to affect workers on a sensory level and to alleviate their fatigued minds and bodies.

It is not surprising that workers subjected to SDO would not find the constant programming particularly jarring, as BGM is already omnipresent in Japan's cities. In this sense, USEN is designing the office to sound like everyday life in sound-saturated Japan. Each channel is matched to the workers' moods, movements, and habits throughout the workday—the music is continuous and sustains them until they leave for the day and are confronted again with BGM in Tokyo's public spaces: sounds of birds chirping on subway platforms to alert visually impaired passengers; a lilting Scottish melody played for pedestrians crossing the street; an easy-listening instrumental version of "Oye Como Va" broadcast throughout a shopping arcade that clashes with the individual stores' BGM; the latest J-pop hit playing in a Family Mart, intermingling with the air conditioning. A worker whose company has subscribed to SDO since its early days confirmed this ubiquity in an interview: "If you think about it, there's music everywhere you go in Japan, whether in the station, crossing the street, etcetera. It feels sort of sad [*sabishii*] if the workplace is the only space without background music."

USEN wants music to be unobtrusive background sound; thus, company discourse minimizes discussion of the act of listening in favor of the music's place within the affective atmosphere of the office. Here, SDO actively encourages a different somatic mode of engagement with the sonic. In contrast to muzak played in North American factories or offices, which stopped workers from developing taste relationships with the music being played, USEN offers company bosses and managers endless choices, endless micro-genres to cater to their workers' moods, tastes, and bodily rhythms. However, similar to Brian Eno's critique of Muzak Corporation, BGM in Japan seems to prevent listeners from listening in a concentrated and critical manner; SDO is a kind of sensory engineering that adds stimulus to the work environment and targets workers on an affective level. Employees use the relaxation channel to deal with the stress of working in confined spaces; they listen to the refreshment channel as a backdrop to lunch, and are transported sonically to imaginary Parisian streets.

Unlike North American white-collar office workers who create individual mood-altering playlists, often drawing on nostalgic music in order to elevate their moods or self-soothe (Anderson 2015, 835), workers who experience SDO are not expected to draw on nostalgic memories in their encounter with BGM, for as USEN conceives of the relationship, "Music provokes memories, whereas BGM is not meant to provoke memories." SDO is engineered so that workers are dissuaded from actively listening to the music because "sustained listening reveries can drift with various degrees of attentiveness from an ambiguous and objectless mood state to the nostalgic unlocking of nests and private memories and associations" (2015, 818–19). USEN's programming is structured to syncopate workers' activities and moods throughout the day without allowing for the possibility of the worker deeply listening, and thus perhaps counteracting the "automatism of cognitive labor" (2015, 819).[35]

Before USEN, most companies played BGM only in their entrances and showrooms, not in their offices. Now, it permeates the workplace: lobby, conference room, break room, and office. One of the main aims of BGM, as a manager told me, is to create a space (*kūkan wo tsukuru*), which implies that sound serves to fill up and fill in office space.[36] He also told me multiple times (echoing USEN's promotional literature) that BGM's purpose is to create a specific atmosphere (*funiki*). For example, Japanese-style restaurants typically feature jazz, but this is only for creating ambience, since, as he pointed out, patrons generally have no idea who the actual musicians are. There are several ways to express atmosphere or ambience in Japanese, and USEN seems to draw on all of these different terms, while often favoring the more abstract term *funiki*, which evokes the emotional ambience and mood of a space.[37] Here, it presupposes an affective atmosphere that does not demand concentrated, active listening. Workers experience music in the SDO program as part of a communal affective atmosphere; listening to SDO is not a private listening experience.

Citing Karl Marx's comments on atmospheres, Ben Anderson points out their necessarily ambiguous and ambivalent status: "on the one hand, atmospheres are real phenomena. They 'envelop' and thus press on a society 'from all sides' with a certain force. On the other, they are not necessarily sensible phenomena" (2009, 78). The ontological indeterminacy inherent to atmospheres evokes the implicit tension between listening and not listening to BGM and the unconscious level of affective atmospheres in offices. "We are not sure whether we should attribute them to the objects or environments from which they proceed or to the subjects who experience them" (Böhme 1993, 114). As atmospheric technique, BGM permeates and enhances space and serves to regulate and manage precarious subjects in their work environments. Paradoxically, USEN prevents workers from consciously recognizing the sound as entertainment, even as it conjures up certain associations for the workers, for example by evoking the sounds of Paris as they eat lunch on their break. SDO promotes a particular type of somatic engagement but also unawareness, as the music during work time is carefully calibrated to achieve this. For instance, the program emphasizes instrumental music and avoids music with Japanese lyrics to prevent workers from listening *too* intently. As we will see, such concerns play out in provocative and sometimes unexpected ways in the business world of sound design and background music in Tokyo.

INTERLUDE III
Retro Shopping Arcades Muzak

In the western Tokyo neighborhood of Nakano, just outside the west exit of the train station, lies the Sun Mall Shopping Arcade (*shōtengai*). The neighborhood of Nakano is the "hip" underground alternative to the more mainstream and touristy Akihabara district, known for its Japanese pop-culture-oriented stores and cafés. Built in 1966 at the peak of Japan's economic climb, just two years after the Tokyo Olympics, the shopping arcade now reflects what is often referred to as Showa retro charm, the style of which is discussed in Chapter 2. The Showa era ended just as the economic bubble was about to burst, leading to the prolonged recession of the 1990s. The period continues to be invoked as an idealized and nostalgic historical period. The consumption of nostalgia is well represented toward the end of the Sun Mall Shopping Arcade, in a section called Nakano Broadway, a four-story, dizzying cluster of vintage shops primarily aimed at collectors and *otaku*, or die-hard fans of Japanese popular culture. One can find used manga, vintage *kajiu* figures, kewpie dolls and doll parts, and railway memorabilia, among other collectors' goods and ephemera.

Part of Sun Mall's nostalgic appeal is also its retro sonic ambience, as piped-in BGM reverberates throughout the length of the structure. Speakers are placed at roughly twenty-foot intervals throughout the length of the arcade's walkways, which are lined with stores and restaurants themselves playing their own BGM. The sonic territory of the stores therefore extends outward, spilling into the main hallway. Small surveillance cameras are placed throughout the length of the Sun Mall Arcade in between each of the speakers, subtly invoking the oft-cited connections between muzak programming and surveillance.

As I walk down the arcade, the BGM fluctuates and increases in intensity when

FIGURE INTERLUDE III.1 BGM speakers in Nakano's Sun Mall Shopping Arcade (*Shōtengai*). Surveillance cameras visible to the right. Photo by the author.

I near each individual speaker. The muzak playing in arcades intermingles, often uncomfortably, with an individual store's foreground music. This sonic dissonance reflects Jonathan Sterne's discussion of the inherent acoustical tensions in the Mall of America, where the background music of the transitional spaces (nondescript, instrumental symphonic arrangements; no vocals) contrasts with the louder foreground music of the individual mall stores. I occasionally stop directly underneath a speaker to hear it more clearly, though this also poses challenges, because I have to avoid disrupting the flow of pedestrian traffic. Standing under the speakers and appearing to be doing nothing, I feel slightly uncomfortable, and perhaps arouse suspicion. I occasionally look up at a speaker, as if this visual confirmation will somehow provide more information about the muzak's source. I decide to track the source of this muzak.

Not knowing where the actual BGM programming might be coming from, I choose the store nearest to me, a women's lingerie shop, and ask a store employee if she has any idea where the muzak is coming from. The young saleswoman does not seem baffled by my query, and she politely assures me that her manager will know—"Please wait just a moment," she tells me before heading back toward the rear of the store. I overhear the faint tones of her conversation with the manager,

FIGURE INTERLUDE III.2
Sign leading toward arcade office.
Photo by the author.

who then briskly steps out from behind the cash register. As she approaches me, she tells me to go to the shopping arcade office (*jimusho*), which happens to be just around the corner and up a flight of stairs. The manager seems confident that the office is responsible for the BGM.

As I walk up the stairs, I notice that there are no windows to see into the office. Not sure what to expect, I knock and slowly enter, apologizing as I tentatively walk inside. The room is windowless and surprisingly free of sound. A young woman gets up and greets me, and I explain my interest in tracking down the shopping arcade's BGM. An older man (the only other employee in the office) who is sitting at the far end of the room looks up when I explain my interest, and he invites me to sit down. Not long after we sit down at a table to discuss the muzak, the young woman brings me a cup of green tea. Before beginning our conversation, I exchange business cards with the office manager. He tells me that the muzak operates automatically from inside the office. It begins at 10:00 a.m. every day and is the same music every day; it never changes. As he says of the sound's repetition, "It's the same, day after day [*mai nichi, mai nichi*] . . ." His voice trails off, and I wonder if I can detect a hint of weariness. Although there is no audible BGM in the arcade office, he explains that the amplified sound from the arcade filters into the office.

I then notice the sound system directly behind me, uncovered and unobscured. Somehow, standing directly in front of the system that is responsible for the arcade's muzak is not quite as noteworthy as I might have thought. The system feels smaller and cruder than I was expecting.

FIGURE INTERLUDE III.3
Music system that supplies the BGM for Sun Mall Shopping Arcade. Photo by the author.

The office manager allows me to take photos and inspect the system more closely. The muzak runs on a continuous loop, unlike earlier Stimulus Progression playlists that built periods of silence into each hour of programming. Because of the arcade's high ceilings and narrow passageways, the BGM works efficiently with the acoustic space, he explains. The muzak is always instrumental music; there are never any vocals. He tells me that occasionally shoppers will stop by the office to ask about specific songs playing in the arcade; they will often mention that they like the muzak, though he explains that "only the older generation seems to appreciate it." Young people hate it, he rues. Perhaps the sound induces a sense of nostalgia for older Japanese, although this question of memory would seem to contradict what a muzak company employee later told me in an interview. For him, the difference between music (*ongaku*) and muzak (BGM) is that music contains the potential to provoke images or nostalgic memories in the listener, whereas BGM does not. Music should be precisely and properly (*kichinto*) listened to. BGM, on the other hand, should just be heard or registered as background. But why, I wonder, should we not expect BGM to contain memories? Is it too disruptive to the commodity form and the process of consumption? Does it unintentionally undermine the consumerism of patrons as they drift through the shopping arcade?

FOUR

Sonic Air Conditioning

Ubiquitous Listening as Mundane Comfort

If Muzak makes people happy . . . and contented in their environment like air-conditioning or a colour-scheme—*How can it not be good?*[1]

Umberto Muscio, President of Muzak Corporation, 1966

ENCOUNTERING BGM

On a recent fieldwork trip to Tokyo, I stayed in what had become my standard lodging, the weekly mansion. In this context, the word "mansion" is a Japanese neologism that refers to architectural materials; it does not evoke the same sense of grandeur as the English word does. The weekly mansion where I tend to stay is a chain budget hotel company—each one looks and feels the same. The buildings' exterior structures vary slightly depending on location, yet the room size and furnishings are standardized, creating a comforting repetition. Like most consumer spaces in Japan, the location where I recently stayed featured background music in the lobby, a common transitional space where BGM is often heard. Sound is smoothing and provides comfort in places that might otherwise provoke anxiety for patrons. The BGM welcomed patrons when they entered the hotel or exited the elevator on their way out, via speakers embedded in the ceiling of the front entrance and lobby. I soon noticed that the music seemed to be the same every time I encountered it—Pachelbel's *Canon in D*, an instantly recognizable and a highly repetitive piece that is often rejected by classical music connoisseurs for being too popular, perhaps because of its utter recognizability and ubiquity. At the same time, the piece is often described as soothing and healing (it is often played at weddings), a kind of "easy-listening" classical music.[2] The piece is well known enough that its presence is not jarring or off-putting to those with no knowledge of classical music who pass through the space in which it is played.

Throughout my stay at the weekly mansion, I entered and exited the lobby at different points in Pachelbel's piece; however, the repetition soon became unavoidable. The sonic repetition melded with the interior design repetition. Occasionally the lobby was silent, which felt somewhat unsettling since I had come to expect the strains of Pachelbel's *Canon* to greet me as I headed out for the day. One day, I spoke with a friendly hotel staff worker and asked her about the lobby music. She told me that it was always classical and, in fact, she added with an overly bright smile, it was always the same song played "over and over." Then, "I'm actually pretty sick of it!" she admitted, the smile eerily never leaving her face.

Sound designers and BGM company employees in Tokyo fill up public spaces like hotel lobbies and retail stores with background music and ambient sound. Sound engineering techniques are described by some sound design companies in highly aesthetic terms, such as painting space (*kūkan*) with sound. Such aesthetic terms, it must be noted, are also often intertwined with terms of control and management. Sound, as one company describes in its corporate literature, can be used to "control the space," evoking long-standing associations, outside of Japan, of ubiquitous background music with social control. For example, in *Noise: The Political Economy of Music*, Jacques Attali examined how music was used to institutionalize the silence of others. Placing particular emphasis on background music, with its linkages to mass consumption and labor, he argued that Muzak served to silence everyday noise through repetition: "Muzak, the American corporation that sells standardized music, presents itself as the 'security system of the 1970s' because it permits use of musical distribution channels for the circulation of orders. The monologue of standardized, stereotyped music accompanies and hems in a daily life in which in reality no one has the right to speak up any more" (1985, 8).

As a form of ambient sound that operates as sonic branding and mood enhancement for stores, BGM is ultimately a form of functional music that creates affective atmospheres through spatial and sensory design, similar to architecture or air conditioning. Foreshadowing the later use of muzak in work and consumer spaces in the 1940s and 1950s, composer Erik Satie posited in 1920 that his notion of furniture music "creates a vibration; it has no other goal; it fills the same role as light and heat—as *comfort* in every form . . . Furniture music for law offices, banks, etc." (cited in Shattuck 1955, 169).

Muzak, described by Jonathan Sterne as signifying "the height of banality," would seem to evoke the low taste by which it is often rejected or condemned. However, with the shift from background music (instrumental covers) to foreground music (original songs) beginning in the 1980s, the very question of taste in muzak and music has been upended. Joseph Lanza, for one, laments the shift in muzak from easy-listening instrumental covers by artists such as Percy Faith, Ray Conniff, and Mantovani, to the use of original Top 40 pop songs—a shift that effectively remakes background music into foreground. Rather than offer customers one elevator music channel, foreground music allows individual businesses to take advantage of the massive selection of in-store music channels, and, in this way, cater to a wide range of tastes. In addition to foreground music's more obtrusive volume, Lanza is critical of corporate sonic branding: "Under this assumption, urbanites might feel more urban when hearing jazz selections at a Starbucks, with the added value of being able to purchase CDs of the same music at the frappuccino checkout. But those with a different idea of the appropriate caffeine concerto are left feeling alienated. And a high-toned New York city bistro might try to slather on a layer of 'class' by piping in classical music, but not everyone likes classical. Classical has even been proven to be a human repellent, such as when New York's Port Authority blasted it to scare away loiterers" (2013, 625).[3]

Much has changed in the past thirty years, not the least of which is Muzak Corporation's steep economic decline. Given the transition from Fordism to post-Fordism, where Muzak's role has shifted from controlling the worker or consumer to the technique of atmospherics (catering to the customer's taste and sensory disposition), it is significant that sound design companies draw on discourses that aestheticize sound design techniques while still evoking themes of controlling space. As a result of this complex history, muzak occupies a tricky, contradictory position in city soundscapes. The right music must be used to create the ideal affective atmosphere, yet its sonic presence is carefully engineered so as not to engage the listener *too* much. As I discovered during my interviews with BGM employees and sound designers in Tokyo, the line between employing music to fill up space and using BGM to evoke feelings of nostalgia and branding (to further commercial aims) is not clear-cut. Muzak's planned invisibility hovers strategically between distracted non-listening (such as classical BGM playing in a hotel lobby or bank) and affective engagement (such as ambient sound in a women's restroom in an upscale department store, designed to enhance the experience).

These various uses of muzak reflect assumptions about the activity of the listener. In his study of programmed music in Minnesota's Mall of America, Jonathan Sterne explains that there is an inherent ambiguity to the figure of the listener, which can refer to the active or passive perception of sound. As he argues, this ambiguity of listening is significant since programmed music "isn't meant for contemplative listening; it also isn't always 'heard' in an entirely passive fashion—rather, it tends to pass in and out of the foreground of a listener's consciousness" (1997, 25).

In this chapter, rather than assume that BGM operates strictly in the background, I linger on the moments when BGM becomes noticed (whether rejected out of annoyance or pleasurably attended to). While most listening is described as either intentional or implicit, BGM is something one passively encounters but does not directly or intentionally listen to; yet BGM *programming* is planned, purposeful, and highly managed. As a sound design company employee put it, "Without a doubt, there's always a purpose [*mokuteki*] with BGM; it's never the case of 'anything goes.'" This project of bringing muzak listening and engineering to the foreground complicates claims of the passivity of background music listening. This fluid engagement with background sound is a significant mode of contemporary auditory experience in the city; some listeners seek out sound-filled spaces in order to relax and tune out, for example those attending classical music cafés (Chapter 2). Other listeners, for example some experimental music patrons (Chapter 1), desire intentional listening experiences that though isolated and unconventional, are nevertheless informed by the city soundscape.

SONIC COMFORT

BGM serves as meaningful sonic material in Tokyo's public spaces and cannot be dismissed as cacophony despite its ubiquity. While much of the "background" music in Tokyo operates as foreground music for sonic corporate branding, BGM includes ambient sounds used to induce comfort and relaxation, as well as instrumental elevator music that intentionally evokes nostalgia for some listeners. The production and engineering of comfort has historically been tied to consumption and the notion of physical comfort as a learned behavior produced through "institutional support and resources," rather than reflecting "innate human traits and dispositions" (Healy 2014, 36). David Bissell, for example, describes comfort as an affective resonance that "moves away from comfort as objectified or as intentional and instead present[s] comfort as a complex set of affective resonances circulated through a variety of tactile, visual and audio media. Comfort is no longer an attribute of an object but more a set of anticipatory affective resonances where the body has the capacity to anticipate and fold through and into the physical sensation of the engineered environment promoted" (2008, 1701).

Throughout much of the sound design corporate literature I discovered both in Japan and in North America, music is frequently categorized in terms such as "optimal" or "appropriate." Sound designers are constantly attuned to the effect of the BGM they create on the overall environment, but more important, on the individual passing through that environment. Background music must be "easy to listen to" (*kikiyasui*) and should not challenge or distract the shopper or worker. Here, we are reminded of the madam at classical music café Mignon expressing worries over patrons requesting her to play *gendai ongaku* (modern music). Even though this music is part of the café's collection, its atonal qualities could be unsettling to other patrons, thus potentially driving away some patrons and ultimately affecting business. Here, it is useful to compare the use of BGM as sonic comfort to ambient engineering techniques such as air conditioning, and the ways in which these technologies produce and inform new sensibilities. In his analysis of the relationship between air conditioning and the act of shopping, Stephen Healy examines how "the condition of 'optimal' thermal comfort delivered by a/c can induce a languid state that can render subjects more open to subjective influence" (2014, 42). He argues that through thermal comfort, shoppers are ultimately made "more vulnerable to enticements to shop" (2014, 37). This is similar to the tactile sensations often associated with comfort, such as an easy chair or a soft pillow.

To compare BGM and air conditioning as ambient technologies begs the question: Does BGM render its listening subjects more "vulnerable to enticements"? Has BGM produced a new auditory sensibility? Setting aside the attacks on muzak as soulless low art, one of the most common critiques of muzak is that it subliminally controls and manipulates listeners; it subconsciously leads those who hear it to shop more and linger longer.[4] This critique implicitly presumes that listeners are vulnerable to sonic intrusions and can be naturally swayed by them. It also presumes the effectiveness of muzak by implying that these putatively manipulative techniques *actually work*.[5] By contrast, in my interviews, sound designers conceded that it is incredibly difficult to track the effectiveness of BGM in retail spaces.[6] While they will occasionally distribute surveys to store customers about the musical selections, most of my interviewees concur that taste is so subjective that it is hard to objectively determine if the BGM programming is working or not. Both air conditioning and BGM share the affective capacity to regularize the environment, yet air conditioning is rarely condemned as a manipulation of public space.[7] Aside from environmental concerns, air conditioning is seen in many quarters as a necessary part of modern life. Muzak, on the other hand, is seen, especially in some parts of Europe and North America, as a noisy intrusion into public space. The anti-muzak movements such as Pipedown in the UK attest to this organized opposition.[8]

BGM is engineered to act on bodies in public and semi-public space. Sound designers concede that people do not actively listen to BGM, yet their programming and design techniques, often informed by strict rules and conventions, would seem to suggest otherwise. They conduct extensive planning and research so that background music and ambient sound generate certain bodily and auditory effects on those passing through BGM-filled spaces. While office workers engage somatically with background music in ways that deemphasize the music itself, so that it enhances productivity (at least, according to USEN's logic), BGM listeners are provoked into certain associations between sound, memory, and genre as they move through public and semi-public space. In all of these sonic practices, background music and sound is meant to do something to listeners, even though it is expected that they are not actively listening.

STUDYING MUZAK: OR, HOW TO WRITE ABOUT SOMETHING YOU'RE NOT SUPPOSED TO NOTICE?

In order to understand the discourses of listening that undergird the logic behind BGM programming, I conducted ethnographic fieldwork with three different Tokyo-based background music companies beginning in 2011. Despite widespread acknowledgment that BGM is not designed to be directly listened to, company employees admitted that it is explicitly designed to have certain effects on consumers and workers. They also told me about receiving positive feedback from store customers concerning their in-store music programming. My fieldwork also revealed that many store workers who are subjected to constant music programming are deeply aware of its sonic presence.

In addition to USEN, I conducted ethnographic research with two sound design companies based in Tokyo: Tōyō Media Links (TML) and Mood Media.

Both of these companies position themselves as high-end sound design firms. They distinguish their corporate identity as antithetical to USEN, in terms of the locations and clients they serve, as well as in their background music programming strategies. Because they frame themselves in terms of "sound design" or "mood," these two companies often described their programming logic in terms of aestheticizing public or semi-public retail space. While they use standard background music genres (jazz, classical, bossa nova, and so on), they also tend to draw on ambient sounds to evoke certain moods, often in spaces that are seen as outside the traditional purview of background music, including bathrooms and store entranceways. All three companies are continuing to expand the physical boundaries of sound programming in Tokyo's public and private spaces, with a clear focus on spaces that did not previously have sound. For example, USEN is now marketing BGM in junior high school hallways, law offices, and medical waiting rooms, while Mood Media and Tōyō Media Links are programming ambient music in women's restrooms and commissioning musical jingles to be played at department store entrances during opening and closing.

Despite BGM's sonic ubiquity, tracking its listening practices proved challenging. Because the BGM listening public is ambiguous, and because those who are subject to it tend not to recognize themselves as listeners, I focused the bulk of my fieldwork on the background music companies' discourses of BGM listening and programming.[9] In order to track and understand the design and programming of BGM and ambient sound, I conducted in-depth interviews with employees (lower level, mid-level, and CEOs) of three of Tokyo's main background music companies. I accompanied sound designers as they visited public spaces (hotels, cafés, bars) that have programmed music. In addition to interviewing BGM employees and analyzing corporate marketing and advertising materials (both print media and online), I also listened closely to the BGM that surrounded me everywhere I went in Tokyo. This was not as simple a process as it would seem. Because this form of music is not to be directly listened to or noticed, I looked suspicious when I stopped underneath built-in speakers in a shopping arcade to listen to the BGM, or when I lingered perhaps a bit too long in an upscale department store restroom with my recorder to capture the ambient sound that was circulating faintly throughout the space.

Trying to make ethnographic sense of muzak in Tokyo thus poses certain methodological challenges. How does one write about something designed to be unheard? Despite the inherent challenges in asking people about their reactions to BGM, a form of sound they were not supposed to consciously notice, I did try to speak to as many store workers as I could. Employees told me that they often tuned it out while at work, though they were eager to detail the genres their store or restaurant featured and when certain genres shifted during business hours. For example, for some Japanese-style noodle shops, classical music is typically played during the day, and jazz at night. As I discovered during fieldwork observing businesses throughout Tokyo and their use of BGM, many restaurants or stores play only one channel. This, I learned through discussions, was often mandated by company headquarters. The workers have little choice of what genres are played and no control over the music's volume. When I asked an employee at a noodle shop about its background music, her immediate response was mild concern: "Do you hate it?" I quickly assured her that I did indeed like it (in this case, mellow instrumental jazz). When I asked if it was supplied by USEN, she responded yes, and quickly confirmed the shop's channel of choice (H-9). The noodle shop worker told me that they wanted the music to be gentle and soothing for the patrons, not overpowering or harsh. The volume was appropriately minimal; patrons could sense the music's presence, but it never inhibited conversations or noodle-making.

I encountered this attachment to a specific channel at other restaurants and cafés too. Many retail employees I spoke with whose employers subscribed to USEN were able to immediately recall their individual selection, based on USEN's myriad choices, each number corresponding to a different genre. Other store employees seemed outright baffled by my interest in something of so little aesthetic value. A store clerk at a 100-yen shop was puzzled and perhaps irked by my query about the origins of his store's BGM, which sounded like classic elevator music (instrumental, jaunty synthesized melodies). His puzzlement told me that BGM should not be commented on, noticed, or recommended. Its sheer unobtrusiveness and anonymity is critical to its putative efficiency.

In contrast to this bewilderment, I found BGM corporate employees and sound designers to be intellectually curious about my ethnographic interest in their sound programming.[10] During my conversations with BGM employees, their position shifted from corporate secrecy—some material they claimed they were not able to reveal—to willful divulging of sound design programming strategies. Very often, formal interviews with BGM employees would organically expand into brainstorming sessions concerning music programming logic more generally. For example, during one interview, as we discussed certain accepted conventions of music programming, one of my sound designer informants was unable to explain why, for example, you would never play bossa nova as café background music at night; it should only be played in the afternoon, he insisted.

It was while interviewing sound design engineers that I began to see intriguing parallels between ethnographic methods and corporate market research. Much of the logic that guides background music programming, especially the creation of moodscapes and affective atmospheres, draws on ethnographic questions concerning the relationship between sound and space and the effect of such sound programming on the individual. Against the rhetoric of background music as unobtrusive acousmatic sound, BGM and sound design companies, like ethnographers, are deeply invested in those moments when BGM becomes noticeable. Not surprisingly, I discovered that BGM company employees were well versed in the English literature (translated into Japanese) on muzak. Because there is no existing scholarship on BGM in Japan in either Japanese or English, I understood their interest in my ethnographic analysis of sound design and music programming in Japan.

As I discovered, most of my contacts in the background music and sound design world in Tokyo have creative backgrounds. Moreover, not only are they well aware of the larger corporate history of muzak as it developed in the West, but many of them are knowledgeable about avant-garde and ambient music, especially in relation to functional music. During most of my interviews, for example, the names of pioneers in sound experimentation, such as John Cage and Brian Eno, or sometimes Erik Satie, came up as we discussed the larger history of muzak. Cage's and Eno's critiques of background music as sonic pacification in the service of capitalism were not lost on my informants. As I noticed throughout my time spent with sound designers (and as they often remarked to me), they not only tended to be attentive toward sound in public space, but were all generally self-described "music fanatics" (*otaku*) and keen listeners of the city, as well as highly attuned to BGM programming trends. During a conversation with a music designer at Mood Media, for example, she told me that she had noticed that McDonald's had recently changed its BGM to something more "stylish."

The highly engineered sonic ambience created by muzak serves as part of the everyday infrastructure of consumer and leisure spaces in urban Japan. The repetition and ubiquity of BGM in contemporary Japan is an effect of the conditions of listening in everyday postindustrial life. According to Anahid Kassabian, we are completely surrounded by music everywhere we go. This omnipresent music "poses as a quality of the environment" and ultimately shapes our listening practices, which are heavily distracted and ubiquitous—we primarily listen while engaged in other activities such as working, cleaning, cooking, commuting, and so on. The ubiquity of much music today complicates our received understandings of music as directly tied to a specific author and specific location (Kassabian 2013, 9). A ubiquitous mode of listening deemphasizes its own status as an intentional auditory practice (2013, 10).

Ubiquitous music's power, then, resides in its sheer pervasiveness. Because we are completely surrounded by music or sound in our everyday lives, it is assumed that we no longer recognize that we are listening to this ubiquitous music. It seems to recede into the background. BGM company employees and sound designers focus on the moments when sound shifts from background to foreground to impact the listener on an auditory and sensorial level. BGM companies constantly seek to develop new techniques for using sound to amplify businesses. BGM is employed to trigger a range of affects and associations, such as store branding, comfort (in the case of genres that are believed to be "easy to listen to," like jazz), nostalgic associations, and occasionally "native" aesthetic concepts of beauty. These techniques are all grounded in stimulating the patrons' sensory experience as they move through the space.

AUDITORY HYPERESTHESIA

Mood Media Japan is a subsidiary of the Canadian company Mood Media, which took over Muzak Corporation in 2011 after it filed for bankruptcy in 2009. Mood Media describes its practices in highly sensorial and aestheticized terms: "The reverberation of sound, images, and scent are various ways to decorate space, similar to designing a space through the use of curtains or paint. This is our vision of experience design." This business ideology emphasizes selling customers a memorable experience or a feeling, rather than specific goods or products. It is predicated on engaging the senses in order to appeal to the customers' emotions: "The more senses an experience engages, the more effective and memorable it can be. Smart shoeshine operators augment the smell of polish with crisp snaps of the cloth, scents and sounds that don't make the shoes any shinier but do make the experience more engaging . . . Similarly, grocery stores pipe bakery smells into the aisles, and some use light and sound to stimulate thunderstorms when misting their produce" (Joseph Pine II and James Gilmore cited in Howes 2005, 290).

David Howes describes this shift toward multisensory marketing as creating a state of "hyperesthesia" in the consumer (2005, 288). After the 1970s, manufacturers became more concerned with "what the *product does* to the consumer," as opposed to what the consumer does with the product (Bijsterveld 2010, 203).[11] While multisensory marketing aims to engage all of the customers' senses, sound plays a prominent role in determining how customers assess certain products. For example, in the case of luxury automobiles, the sonic experience of customers is precisely engineered to appeal to their associations between sound and luxury brands. The sounds of the car's engine and exhaust are tied to the image of specific car brands. BMW, for example, has hired acoustic engineers to ensure "that the engine sounds sporty, the doors produce a reassuring solid sound . . . [and] the buttons click for a good reason" (Daniel Jackson cited in Bijsterveld 2010, 201. In this case, sound not only evokes a specific brand image (here, "sporty"), but also conveys that the car is structurally sound and safe. It is not surprising that with the shift toward hybrid, and thus less noisy cars, some patrons have associated noiselessness with dangerousness, a feeling that something is off. Tesla Motors discovered that some customers felt their electric cars were *too* quiet, a reaction that led an engineer to design software that would provide a range of engine sounds as a kind of sonic reassurance (2010, 201). This apparent discomfort with noiselessness, especially in situations where we expect noise, is a sentiment that some BGM companies in Japan, especially USEN in its SDO program, draw on in order to appeal to workers. Spaces that are believed to be without sound are often described by BGM designers as being "lonely" or "unsettling." Thus, amplifying continuous background music not only will comfort stressed workers, but also can enhance productivity.

Mood Media Japan focuses in particular on using sound as part of experience design. Background music and ambient sound will "decorate" the space and enhance the overall sensorial experience for the patron. I interviewed the CEO of Mood Media as well as one of the company's primary music designers, and we discussed the ways in which BGM has become standardized throughout public space in urban Japan. Jazz, for example, is one of the most commonly heard forms of BGM throughout certain types of Japanese-style (*washoku*) restaurants, as well as bars and pubs.[12] As the CEO explained, "Jazz is seen as modern and high class," whereas other genres, especially older Japanese musical forms such as *enka*, are seen as outdated and retro. Many of the sound designers I spoke with were frustrated by such genre conventions, claiming them to be unoriginal and uncreative.

The music designer at Mood Media expressed a desire to hear unexpected BGM, admitting that "when I go to a really hip *washoku* restaurant and hear either jazz or *koto* [traditional Japanese stringed instrument], I find it really uninspiring." She contrasted this with a surprising but pleasurable encounter with BGM in a fashionable Tokyo neighborhood: "I went to a Mexican restaurant in Omotesando once, and the BGM seemed like it was personally selected by the staff. They played a range of styles that seemed to be on shuffle, then all of a sudden there was *enka*! It seemed very Tarantino-esque; it reminded me of *Kill Bill*. I thought it was very cool, but extremely unusual." She laughed, "I was really surprised."[13]

The designer later took me on a tour of hotels and restaurants where Mood Media has programmed background music and ambient sound. During this tour, she explained the programming logic for each space, which included hotel lobbies, elevators, bathrooms, restaurants, bars, and private rooms. We spent a fair amount of time at the Tokyo Palace Hotel, a four-star hotel in downtown Tokyo that caters to wealthy business people.[14] Unlike the repetitive BGM in the lobby of the weekly mansion (a chain hotel, it should be reiterated), the background music in the Palace Hotel was texturally varied, as each space within the hotel offered a distinct sonic environment, including the lobby, elevator, hallways, and bathrooms, among other spaces. This background music technique marked the hotel as a non-place trying to assert itself otherwise.

Mood Media's music designer explained the logics of sound design as we walked through the hotel together. Complicating the association of BGM with the mundane, she told me that she drew on Japanese aesthetic concepts of silence and beauty to inform her programming design. In order to solidify the hotel's theme of "beautiful Japan," she was inspired by traditional sounds of nature: "When I was selecting the music for the hotel, I used the sound of *shishi odoshi* as a motif." A bamboo water fountain often found in Japanese gardens, this device is used to scare away animals such as deer or boars. It produces gurgling and calming sounds of water flowing through a bamboo tube; the water flowing through the tube eventually shifts the center of gravity, thus tipping the bamboo tube back and hitting the rock against which it is resting, ultimately producing a sudden clapping sound.[15] She described the moment right after the bamboo tube is completely filled with water and the clapping sound resonates as a moment of silence or calmness (*isshun no seijyaku*). She framed it in culturalist terms, as a "fleeting sense of beauty" highly valued in Japanese culture. This moment, she went on to explain, refers to *ma*, an aesthetic concept that relates to silence or space, a sense of being in-between. "It is the time between events, the space between objects, the relationship between people, or that moment in a person's mind between thoughts. It is the white space in a pen-and-ink drawing, the pause between notes" (Galliano 2002, 14).

When used in the context of a musical performance, *ma* can signify a positive appraisal by suggesting that the "individual sounds of the music did not occur in an empty vacuum and that the silences in the music were full of a sense of 'betweenness,' or *ma*, for *ma* reflects aesthetic sensitivity" (Galliano 2002, 15).[16] This was the first time I had heard BGM programming being explicitly articulated in cultural terms. In contrast, communication scholar Gary Gumpert has characterized muzak as "amniotic fluid"—sound that constantly surrounds us and is continuous, repetitive, and calming. One of the managers at USEN told me that the company's audio systems are engineered such that the volume between song transitions is constant to avoid jolting the listener. Transitions between tracks are smooth and uninterrupted. Mood Media's sound design for the Palace Hotel complicates these assumptions about BGM as providing a continuous sonic blanket by being tailored to the spaces between sounds, something one might expect to be anathema to the very concept of BGM.

Like Mood Media, Tōyō Media Links is a small but prominent sound design company that manages sound in a variety of what it considers to be "high-end" retail spaces, including various spaces within department stores (restrooms and entranceways), upscale hotels, medical clinics, and banks. When I contacted the sound design team at TML to arrange an interview, one of the members asked for permission to also conduct a "reverse interview." When we met, the team members explained that it was rare for them to be interviewed by someone who is not either a journalist or a business person, so they wanted to learn more about my research. I met with a sound design team member, customer support team member, and the customer support team leader.

Founded in 1962, TML does more than just sound design: it also offers surveillance camera services, digital signage, and more recently, scent programming. Scents are often intertwined with sound to create a rich, multilayered atmosphere of sensory branding. Unlike USEN's customers, both retail and private, TML's customers are unable to select specific music channels because the company retains full control over programming. The right music for the right environment is ensured by extensive consultations with individual clients to determine the appropriate fit. As staff members told me, they will first meet with clients to understand the concept they have for their space, after which the sound design team will recommend specific music programs. Though BGM is highly engineered, TML's workers do not want people to overtly notice or listen to it in either consumer or work spaces (where they may be distracted from work or consumption); the focus is on creating an appropriate affective atmosphere and sonic branding.

One of the sound design team members revealed specific rules concerning BGM programming, including that BGM songs cannot have Japanese lyrics, as that would be too distracting. Songs in English are deemed acceptable, however, since it is unlikely that people will notice these. We discussed conventions regarding specific genres tied to specific settings and times of day. Bossa nova, for example, is one of the more common genres to hear in retail settings (especially cafés). But it is only played during the afternoon.[17] To hear bossa nova at night in a café would feel unsettling (*iwakan*), as one employee explained, even though he was unable to describe why this would be so, other than the fact that it defies conventional use of BGM. He further noted that although it would not be unusual to listen to bossa nova by oneself at night at home, it would be very unusual to hear bossa nova playing in a bar or lounge at night. In the background music world of Tokyo, bossa nova is used solely in the afternoon. These genre conventions were explained as standard procedure in Japan. "It's always bossa in the afternoon and jazz at night," a sound designer neatly summed up.

Other fixed conventions of genre include the widespread use of jazz as BGM, especially in Japanese-style restaurants.[18] When I spoke with the sound designers, they agreed that jazz is easy to listen to and can fit with any type of restaurant. In terms of its sonic properties, jazz is considered "safe" and comfortable for the patrons (*ochitsuita kanji*); it is not difficult listening, as *onkyō* is. Many sound design employees I spoke with articulated BGM engineering conventions as something particular to Japan, and playing traditional Japanese music or instruments was similarly guided by certain widely accepted conventions. For example, it would not be unusual to hear traditional Japanese instruments like the *koto* as background music in a restaurant, yet the popular ballad form of *enka* would never be used as BGM. The message conveyed by such emotionally weighty music, as a sound designer explained, is too strong for the listeners. Here, "too strong" indicates that restaurant patrons might be distracted by the nostalgic lyrics (in Japanese) or the music itself, steeped in themes of longing, pain, and unrequited love.[19] Distracted by *enka*'s lyrics, characterized by Christine Yano as expressing "clichés of excess" (2002, 90), patrons might then unwittingly *listen* to the music. *Enka* is also tied to a particular moment in Japanese history, the Showa era, and is often seen as a retro (if quaint) musical form, whereas traditional instruments such as the *koto* operate outside of history, denoting a more generic sonic property of so-called Japaneseness. As sound engineers explained to me, the sound of *enka* evokes an older image, whereas newer (often instrumental) musical genres like jazz evoke feelings of comfort, ease, and coolness without burdening the (Japanese) listener with the weight of things that are out of time.

TML stands out from some of the major BGM companies due to its highly aestheticized approach to sound design. Its corporate literature describing the company's approach to background music programming is peppered with poetic language about the powerful role of sound in space. TML defines sound design as "decorating space with sound" as well as "coloring" space with sound. The ability of sound to beautify space is intertwined with themes of control and management. BGM, as TML's literature describes it, allows individuals, clients, and companies to control space (*kūkan*). During my interview with some of TML's sound design team members, they compared BGM to other infrastructural elements of space, such as air conditioning and temperature. TML company literature frames BGM as something that will improve business (*shinka dekiru*), a discourse that invokes the techno-utopic foundations out of which the company emerged and to which it is still indebted. From its early days when it was used to stimulate factory workers to its function as audio architecture and environmental design, Muzak has always evoked progress and modernity (Vanel 2013, 48–49).

Although its corporate history is similar to that of USEN, having first emerged in the 1960s, TML frequently positions itself as distinct from other BGM providers in terms of corporate strategies, ethos, and clientele. It offers more intense guidance to its clients in order to determine the appropriate BGM for a particular retail space. While USEN offers pre-selected programs and packages for a massive range of customers—corporate and private—TML curates an entire playlist for a store, individually tailored to that store alone. As a result, its service fees are much higher than USEN's, and it typically works with more selective high-end clients. USEN, by contrast, sells BGM on a mass commercial level, which includes convenience stores, fast-food chains, and 100-yen shops. Like most other BGM companies, TML does not typically compose the music or have access to the recordings that make up the playlist, and therefore must rent songs from a recording company. Clients are thus paying not for music, but for TML's sound design employees' expert knowledge in choosing the right background music for the specific environments.

TML, like many other BGM companies, often programs music in transitional spaces such as store entrances, hallways, hotel lobbies, and restrooms. Two of its recent programming contracts stood out to me: programming ambient music in women's restrooms in NEWoMan, the newly opened department store in Tokyo aimed at women; and commissioning music to be played at the entrances to this same department store. While public restrooms globally have become saturated with background music in recent years, it tends to be foreground pop music, often tied to the store's brand identity. The most common sonic landscape of public women's restrooms in Japan is that of white noise, due to the widely disseminated "sound princess" (*oto hime*), a white-noise device mounted inside individual restroom stalls. The *oto hime* can be heard in restrooms in department stores, train stations, and businesses, among other spaces.

First invented in 1988 by the Toto Corporation (internationally famous for its technologically advanced toilets), the sound princess was designed as a means to counteract the wasteful running of water in order to mask bodily sounds in public restrooms.[20] The device thus draws on the power of background sound and ambient noise as a sonic blanket to create privacy between strangers in cramped public spaces. The impetus for the sound princess is steeped in culturalist and gendered discourses concerning Japanese women's so-called tendency toward embarrassment at the prospect of other people hearing them in public restrooms. According to Toto, the sound princess was designed to combat women's supposedly "delicate mentality," a logic that is often explained in environmental terms. In this case, the wasteful running of water is replaced by the technologically mediated sound of trickling water that can be controlled by the individual in her private stall. The sound of the water itself is aestheticized, as the device offers users two types of sound to choose from: a murmuring lake or a murmuring lake with birds chirping. The company claims that such nature sounds provide a "relaxing effect."

TML was contracted to produce BGM in the women's restroom in the brand-new department store NEWoMan, managed by the Lumine department store chain, which opened in the Shinjuku neighborhood of Tokyo in 2016.[21] Aimed at "mature women who are looking for truly fine items," NEWoMan offers a range of stores from upscale fashion boutiques and shoe stores to restaurants, French bakeries, and cafés. It also features a rooftop garden, child-care center, and full-service women's health clinic. As the TML employees pointed out to me, public restrooms are incredibly important spaces in Japan, and their cleanliness is often reviewed in online sites such as Tabelog. This applies especially to restrooms located within department stores, which are typically connected to railway stations and thus used frequently by commuters who do not actually shop in them. As the employees explained during our interview, the more positive experience someone has in a public restroom increases the likelihood that they will associate that positive atmosphere with the department store itself. Ideally, satisfied washroom-goers will become devoted customers.

It is for this reason that women's restrooms are very carefully and strategically designed. During my interview with TML employees, they pointed out that men's restrooms are typically small, cramped, and narrow, while women's restrooms are much more spacious and comfortable, often resembling lounges with seating, plants and lush greenery, and now, background ambient sound. While the sound designers at TML bemoaned the fact that online reviews rarely comment on the background sound (not surprisingly, reviews tend to focus on the bathroom's cleanliness and scent), they reiterated the importance of women's restrooms as a space of comfort, ease, and, not insignificantly, consumption. Some department store restrooms will feature internal windows with goods on display, thus making the restroom function as an extension of the store itself. Sound, in this context, is thought to solidify consumer loyalty and create an atmosphere of calm. In these toilet settings, the sound engineers explained to me, the BGM must be ambient music, and cannot have a discernible melody or rhythm, or be too "energetic." As we discussed the logic behind using ambient music in a space like a restroom, one of the sound designers jokingly mentioned Brian Eno, and we all laughed at the absurdity of Eno's concept being applied to a department store restroom. In fact, as discussed in Chapter 1, Eno's theory of "ambient music," a form of music that "is intended to induce calm and a space to think," was vehemently opposed to the commercial thrust of the Muzak industry, where music was programmed solely to stimulate consumption.

In some ways, TML's programming at NEWoMan draws on some of the key features of Eno's ambient music: it should be as "ignorable as it is interesting," and it should accommodate "many levels of listening attention without enforcing one in particular." Of course, where this association falls apart is when we consider that this BGM programming is designed to solidify a department store's brand identity, something that Eno's notion of ambient music directly rejected.

TML notes in its press materials for its NEWoMan bathroom design that the lush "display of plants is so beautiful that it makes you forget that you are in a restroom." The restroom's atmosphere is described as a highly sensorial and branded experience in which the greenery display, stylish hand soap and other amenities, and Daisō brand hand dryers work together to create a "refined atmosphere" (*senren sareta funiki*). As TML employees explained, the ambient BGM of "relaxing sounds" was chosen to "perfectly match" this restroom's space.

One of the customer support team members for TML spoke approvingly of the bathroom's ambient BGM, noting that it was very "cozy and comfortable."

Drawing on the widely accepted practice in Japan of using specific melodies to indicate that stores (especially department stores) are closing and opening, TML offers services it refers to as "sound logos" and "welcome sounds." Sound logos are described as effective ways to transmit campaign images, while welcome sound is framed as symbolic sound playing at store entrances to greet customers. In 2016, TML commissioned Tokyo-based jazz saxophonist Kikuchi Naruyoshi to compose a melody for the opening and closing jingle of NEWoMan. In its advertising brochure, TML describes how it selected music that would be fitting for the store's concept, "new experiences in a new city for new women."

In its press release announcing the sonic vision for NEWoMan, TML stressed that the three-minute version of this melody is heard when the store opens (at the entrance), but the full version can be heard only when the store closes, and referred to it as a "rare recording." The press release urged listeners to come check out this "hip" melody in person. TML employees told me that occasionally, fans of Kikuchi will specifically come to the store right at closing so they can hear this (very short and fleeting) music, and will often document and record it.[22]

By featuring a fairly well-known jazz musician, TML's use of BGM complicates received understandings of muzak as unobtrusive, anonymous, background sound. In this case, patrons (fans of the artist) are deliberately seeking out this BGM, not as background listening. Many shoppers, the sound designers admitted, will not recognize the specific artist commissioned to perform the store's opening and closing melody, thus enabling it to operate strictly as BGM—anonymous background music. At the same time, devoted fans of an artist will seek out his or her music to hear and capture through recordings. For them, it is the antithesis of BGM. The rarity of this "sought-out BGM," something that can be heard only at the store entrance and not purchased or downloaded as a recording, adds to the allure of hearing it "live." Similar to Mood Media's music designer drawing on the space between sounds to inform her sound design, TML also approaches BGM engineering in ways that challenge received understandings of muzak as formulaic, mundane, and widely applicable across a range of environments. In contrast to TML's curated site-specific offerings for individual clients, USEN, the most well-known purveyor of BGM in Japan, offers its customers, whether private or retail, the ability to select their own personal BGM.

I asked a manager at USEN to describe the differences in listening modes between music proper (*ongaku*) and BGM, and he informed me that "music is something you listen to properly, while BGM just enters your ears." Interestingly, the term he used in Japanese to refer to the process of music listening, *chanto*, indicates a normative sense of listening, a process that demands labor and intentionality on the part of the listener; this is the supposed antithesis of background listening, which demands no concentration from the listener. BGM listening, for this manager, should not be framed as passive non-listening, but instead as a meaningful mode of auditory engagement. Sound, in fact, indicates that business is running smoothly.

USEN's music programming is spread widely across retail and now white-collar office settings throughout Japan. Unlike some other sound design companies, USEN also provides a household-centered vision of BGM that it markets as creating a more pleasant and comfortable domestic environment through total immersion in sound. In this program, BGM is ubiquitous and embedded in speakers throughout the home for collective family listening, while also allowing for the creation of individual (and often highly gendered) private moodscapes, such as the "Home Spa" channel, which is marketed as a channel for mothers. The presence of BGM in the household, according to USEN's promotional literature, will strengthen mutual contact (*fureai*, or "touching-together") between family members. This logic is similar to that which guides background music programming for office workers: to strengthen social and affective ties.[23] Private and retail customers are offered the same wide variety of channels, which are broken down into very specific micro-genres, including slow jazz, light music, pachinko music, healing music, morning music, and Swiss yodel music. In addition, BGM itself is further refined and categorized into a variety of distinct genres, including "*Mukokuseki* BGM" (nationless BGM), nostalgic BGM, seaside BGM, resort BGM, and Chinese BGM.

For the home listener, BGM serves to create a mood or atmosphere conducive to lounging or relaxing with family. Similar to Spotify's offer of "playlists for every mood" for its listeners, USEN's private customers can choose channels through a search engine on the company's main website that allows them to search channels based on mood (*kibun*) or scene: moods range from "the feeling of a home café" to "times when I want to cry"; artists range from Japanese *enka* diva Misora Hibari to Mariah Carey.[24] Paul Allen Anderson describes how the process of streaming music from such mood-based playlists reduces "the anxiety of excessive choice and retains the element of music discovery: the user need not even name a particular artist, only the mood they want to mimetically represent or the activity the music will accompany" (2015, 827–28).

Despite the widely accepted genre conventions for BGM in many public spaces, especially the restrictions against music with Japanese lyrics as too distracting, or certain genres being limited to certain times of the day, BGM for the home listener is explicitly marketed toward personal mood management and does not contain such restrictions. In fact, private listeners are encouraged to draw on nostalgic associations with music, including music in Japanese, in order to heighten their emotional connection to the music.

The manager at USEN I interviewed referred to the same genre conventions described by the sound designers at TML and Mood Media, especially the standard use of jazz, and expressed frustration with these conventions. As he explained exasperatedly, such rules become codified and widespread because companies follow these practices and copy each other, without trying anything new. These static programming sensibilities are frustrating, he said, because music itself is dynamic and constantly changing. Determining that a specific type of music works in a particular type of space or at a particular time of day is a form of marketing, he explained, but it is also a way of creating a new market. Referencing my discussion with TML sound designers regarding jazz as the most common musical genre in Japanese-style restaurants, the manager lamented a lack of creativity and originality with BGM programming in restaurants and bars. He expressed a desire to radically reimagine the relationship between music and (retail) space, jokingly telling me that BGM providers should try opera instead of jazz.[25] He then mentioned that the company is seeking to create new markets for BGM targeted at tourists, for example in hotel lobbies and capsule hotels for the Tokyo Olympics in 2020.

In an unexpected admission, the manager later confessed that "it would actually be fine to have no background music in Japanese-style restaurants because the natural soundscape could serve as a kind of BGM. The sounds of chopping and food being prepared is in itself a kind of music. But of course," he admitted laughingly, "we would be out of business!" This attunement toward ambient sounds that would not typically be considered musical is in line with what some of the listeners at Off Site (see Chapter 1) told me about the impact of city sounds on their auditory experience inside the performance space. Some listeners described how they became more attuned to the city soundscape after attending shows at Off Site, a music venue with unusual sound restrictions where outside sounds regularly filter into the space and intermingle with the live performances.

BGM AND MUNDANITY

The sound design companies that I spoke with often stressed to me that their company was radically distinct from USEN. Despite their diminutive size and impact relative to USEN's domestic wide reach, they seemed to relish their aesthetic distinction from the biggest BGM provider in Japan. As they told me, their clients rely on their expertise to curate and select the appropriate background music for their companies. Thus, their services are more expensive than USEN's. Beyond the business strategies, sound design companies also seek to distinguish themselves from USEN on aesthetic grounds. For instance, employees often pointed out to me that they served more "high-end" clients, a strategy that highlights a sense of corporate sonic exclusivity. One sound design company jokingly referred to USEN's BGM style as "ready-made goods." The underlying assumption is that the services of sound design companies are site specific to each individual retail environment, rather than USEN's generic mass-marketing focus. There is also an implication that the BGM is of a higher quality, personally selected and curated by an expert rather than "ready-made" banal music.

The mundanity implicit in muzak was taken to new marketing heights in the 1980s by the Japanese chain store Mujirushi Ryōhin. Shortened to Muji, the full name translates to "no brand quality goods." Indeed, Muji's overall marketing strategy was precisely to offer basic and emphatically unostentatious goods.[26] Muji's chief advisor Tanaka Ikko described this corporate philosophy: "You may feel embarrassed if the person sitting next to you on the train is wearing the same clothes as you. If they are jeans, however, you wouldn't be worried, because jeans are what we could describe as 'basic' clothing. All Muji products are such 'basic' products" (cited in Holloway and Hones 2007, 557). Of course, in its project of branding no-brand products, Muji paradoxically ended up solidifying its own unique style, a paradox that can also be applied to USEN-sponsored BGM. The more closely one listens to background music in Tokyo's public space and talks to those who are regularly exposed to it, such as store workers, the easier it is to determine whether BGM is USEN-sponsored or not.

The first Muji store opened in the Aoyama neighborhood in Tokyo in 1983, the same year Muji commissioned techno-pop musician Hosono Haruomi to compose in-store BGM. The cassette recording titled "Watering a Flower" consisted of three tracks, yet only one was used in the store.[27] To intentionally listen to this BGM from today's perspective—as foreground music—can be a pleasantly surprising experience. The repetition is subtle and never feels insistent or overbearing, as is the case with much contemporary BGM. There are occasional flashes of atonality, but they never feel jarring. "Watering a Flower" gestures to the aesthetic sensibilities of Eno's 1978 album *Music for Airports*, because the listener can easily enter the music and listen, or just as easily tune it out and play it in the background. Roughly twenty years after it first opened, Muji returned to BGM by releasing a series of in-store CDs. In contrast to the early avant-garde-inspired work as commissioned by Hosono, this series takes a multicultural approach, with each recording in the series offering (often stereotypical) musical soundscapes from a different region in the world, including Paris, Sicily, Ireland, Buenos Aires, Hawaii, Stockholm, and Rio de Janeiro.[28] Muji describes the BGM in this series as the "background music to life."

Muji markets its clothes as "basic," suggesting that its products are "designed to blend into non-Muji environments" (Holloway and Hones 2007, 558). This nod to visual anonymity and seamlessness with the environment perfectly aligns with BGM logic, which stipulates that music should remain unobtrusive as it facilitates the relationship between humans and their environment. The descriptions of Muji products in its store catalogues suggest qualities that could easily be applied to muzak: "discreet, muted, transparent, understated, unobtrusive, and unostentatious" (2007, 558). It is precisely this quality of BGM that is appealing to some listeners who see the banality and seeming lack of authorship as evoking a sense of stylish sonic anonymity.

BGM AS ANONYMOUS POP MUSIC

During a summer research trip to Tokyo in 2010, I stumbled across a CD release at Tower Records, Shinjuku, in the avant-pop section. This marketing category features a wide swath of genres, including experimental music (noise music is included here), modern music, ambient, exotica, and miscellaneous genres such as sound effects compilations. This particular CD caught my eye, for it was displayed very prominently with a handwritten write-up by a store employee that highly recommended the recording. Moreover, a cleverly staged visual parody of supermarket fliers, the cover image awash in soft pastels, jumped out at me. Featuring a young man and woman posing in generic clothes advertised as "men's and women's casual wear," the cover indicates that the item is "half off," and the discount will be taken off at the register.[29] The back cover continues the theme

FIGURE 4.1 Front cover of *Music for Supermarket* CD (2007), Sweet Dreams Press. Image courtesy of Fukuda Norio of Sweet Dreams Press.

of the supermarket as a nonthreatening and generic space of consumption. The track listing is featured in the upper left-hand corner against a washed-out and shadowless pale mauve backdrop that evokes an ambiguous time period. Occupying the majority of the frame, however, is a staging of banal consumer goods: a straw handbag is flanked by two bright yellow lemons and a French baguette, a fluffy bouquet of baby's breath, and women's sandals that might be considered to be targeting middle-aged female customers.[30]

Composed and produced by musician Hosono Shinichi (formerly of J-pop bands including Magoo Swim, Sunny Day Service, and Puffy) and artist and graphic designer Odajima Hitoshi, *Music for Supermarket* (2007) was the first album release by Sweet Dreams Press, an independent record label that also produced a regular music magazine. On its website, Sweet Dreams explains that its employees painstakingly composed original songs to sound exactly like BGM one would hear in a supermarket. The album liner notes describe the genealogy of the project, tracing it from Muzak's familiar historical lineage beginning with John Cage and moving through to Brian Eno. In fact, the notes mention explicitly that the album was released twenty-seven years after Brian Eno first came up with the concept of ambient music while convalescing in the hospital. The album title playfully evokes Eno's legendary *Music for Airports* album.

Though the album is described in the liner notes as "fake easy listening," Sweet Dreams expresses some hesitation with labeling its recording as a complete parody. It acknowledges the standard critiques of muzak as "canned music" and "valueless and soulless music," while also referring to muzak's listeners as zombies. This reference comes from George A. Romero's classic 1978 horror film *Dawn of the Dead*, which takes place inside a mall where muzak plays a prominent, if at times parodic, sonic backdrop.[31] Intriguingly, Sweet Dreams specifically calls out BGM listeners in the liner notes, explaining that their engagement can be compared to a "feeling of déjà vu." As the company writes in the notes, "even if one is hearing the track for the first time, it feels as if you've heard it before."

Music for Supermarket sonically matches the parody artwork displayed on the album's covers. The music is deliberately lo-fi and invigorating (*sawayaka ni*); the liner notes state that a Roland SC-88 synthesizer (from 1994) was used, which lends the album a retro yet relentlessly cheerful tone.[32] Indeed, this lively mood continues throughout the album with song titles ranging from "Summer Clearance Sale" to "Afternoon Hills" to "Tropical Island." The individual tracks vary in tone and sentiment, and some occasionally veer into minor keys. However, the overall mood of the album is free of tension. Aside from the store closing track, the individual tracks are short enough that the listener never feels burdened by the weight of repetition, often seen as the primary characteristic of muzak. The penultimate track draws on the widely used convention for stores, often department stores, to broadcast a voice announcement stating that the store is about to close. This announcement, in a smooth female voice, repeats in formal Japanese multiple times over a backdrop of synthesized strings. "Thank you for visiting our store today. The store is about to close. We look forward to seeing you again. Thank you very much." Repeat ad nauseam.

To keen listeners of muzak, *Music for Supermarket* gestures to an older era of what is sometimes disparagingly referred to as elevator music. It is a specific style, rarely heard in North America after the rise of foreground music in the 1980s. But it is a style that is still often heard in Japanese supermarkets, 100-yen shops, shopping arcades, and convenience stores. While it would be easy to read this recording simply as an ironic take on supermarket muzak (and to be sure, there is a winking sense of parody), the liner notes force the listener to engage with questions about the ubiquitous (and presumably anonymous) pop music that surrounds them as they shop and consume. As the label owner of Sweet Dreams, Fukuda Norio, explained to me, the album serves as a nostalgic homage to retro BGM, especially the USEN-sponsored BGM that was heard in

FIGURE 4.2 Soy sauce music. Interior of CD jacket portraying album producers as generic supermarket employees stacking the shelves with soy sauce and miso. Hosono Shinichi (left); Odajima Hitoshi (right). Image courtesy of Fukuda Norio of Sweet Dreams Press.

supermarkets.[33] Paul Allen Anderson describes the appeal for some aesthetes who take pleasure in old-school Muzak "precisely because of their mechanically exteriorized emotional content, alienation effects, and desubjectivizing quality" (2015, 819). I spoke with Fukuda and Odajima, the main graphic designer involved in *Music for Supermarket*, and we discussed how part of the appeal of BGM is its anonymity, despite its utter ubiquity: "It's exciting because no one is listening to it." Similar to how the sound designers I met with were astute listeners of the city, the musicians and artists behind *Music for Supermarket* revealed a similar fascination with the BGM that surrounds them on a daily basis.

CONCLUSION

Tokyo Listening, Listening to Tokyo

Immediately following the March 2011 Fukushima earthquake and nuclear disaster, calls for self-restraint (*jishuku*) circulated in the news and popular media in Japan. The call to restraint focused mainly on domestic energy consumption, but in some cases it also extended to *sonic* restraint and austerity in the areas outside Fukushima as a show of national solidarity with the devastated region. As a result, Tokyo politicians campaigned without the usual blasting of megaphones, stores turned down piercing electronic melodies, store barkers were silenced, and school bands momentarily stopped playing (Belson and Onishi 2011). The discourse of *jishuku* was couched in the rhetoric of collective mourning, thus the calls for austerity implicitly evoked the idea of urban noise as pleasurable or celebratory. This sense of pleasure is also, of course, tied to noise's crucial role in advertising and selling products. City noises were to be muted and put on hold in order to properly mourn, the call for restraint suggested. Neither shopping nor its phantasmagoric sonic shadow were to be celebrated.[1]

Although this sonic austerity was fleeting, its impetus helps reveal the logic of the city's sonic environment. The assumptions about sound and city noise that informed this momentary call for restraint post-Fukushima undergird the central premise of *Tokyo Listening*, which is that the individual's relationship to the city is an embodied and sensory experience through which sound orients people's movements, affects, labor practices, and consumption. Sound design companies' framing of BGM-less spaces as "lonely" or "unsettling," however, takes on a different valence in the immediate post-Fukushima context. An absence of sound suggests the absence of life. Sound reflects normalcy, life returning to an everyday routine.

As I write this conclusion amid the white noise of a fan and music playing softly in the background at home, I am reminded of the manager at USEN rejecting the critique of sound-saturated societies as something aberrant and nonsensical. His defense of a sound-saturated society suggests more than simply an employee defending his business's corporate philosophy. The sonic practices discussed in this book cannot be solely reduced to the use of sound as corporate branding and manipulation. Sound is also deeply tied to the creation of comfort, happiness, and relaxation for city residents: the unexpected pleasures of encountering nostalgic music wafting through the shopping arcade; the café madam playing a recording based on patrons' unique musical tastes; a white-noise machine in a restroom stall providing the feeling of a protective sonic blanket.

The listening practices explored in this book raise questions that resonate beyond a specific geographic region. The Tokyo-based sound designers' framing of spaces without sound as lonely or uncomfortable is indicative of global trends toward the increasing use of soundscaping technologies—including white-noise apps—to either create privacy in public space or provide a comforting sonic backdrop while alone. The use of sound, especially environmental sounds, as a tool for mood management is not a recent trend, nor is it unique to Japan.[2] Office workers in Tokyo are surrounded by BGM throughout the workday, including nature sounds, in an effort by their managers to make them more productive and happy workers; this practice gestures to the use of private mood-regulating playlists by many office workers or creative laborers in North America. Yet what it means to feel calm or be a more productive citizen varies widely, and sound is deployed in varied and often contradictory contexts.

As a way of negotiating Tokyo's dense sonic environment, some city dwellers seek out shared listening spaces such as music cafés or experimental music venues. Sound operates very differently here, as compared with how BGM is used to make white-collar workers more efficient and attached to their work. In the music café, patrons rely on high-volume analog sound (rather than seeking out quiet and private solitude) to relax and tune out the city's deafening soundscape. Café sounds do not serve as the auditory backdrop for work or becoming more efficient laborers, but rather for drifting between varying states of attunement, moving from distracted to more concentrated listening. While many Tokyo residents may ignore or shut out the constant BGM surrounding them, there are others who see this sonic "amniotic fluid" as providing a sense of comfort, an assurance that things are running smoothly.[3]

NOTES

INTRODUCTION Listening to the City

1. In this book, I capitalize muzak when I am referring explicitly to the Muzak Corporation and use the lowercase when I am speaking of the muzak phenomenon more broadly. I generally use the Japanese term for muzak, "BGM" (background music), throughout the book when I discuss how background music is consumed and produced in Tokyo.

2. See Bull (2007) for a rich analysis of mobile iPod listening that explores the impulse of the urban dweller to sonically disengage from the sounds of the city in order to create a private "auditory bubble."

3. See Peterson for an ethnographic analysis of the public concert audience in Los Angeles, described as "constituted as a collective, affective musical public and a civic moral community through embodied experiences of listening, dancing, and feeling moved by the music" (2010, 15). While the listening community I examine is more diffuse than the one Peterson analyzes, I am nevertheless concerned with tracking collective listening experiences throughout Tokyo's urban spaces.

4. See Hankins and Stevens (2014, 1) for discussion of the debate over "sound leakers" in Japanese public transportation.

5. The territorializing power of sound to demarcate and carve up space is the subject of a sensory history by Alain Corbin concerning the sonic and cultural presence of church bells in the nineteenth-century French countryside (1998). In this case, the sound of the bells sonically and spatially signified those who were part of this auditory community; those who were within earshot of the bell's reverberation were sonically marked as members of the community.

6. In Chapter 3, I examine how nature sounds such as birds twittering are brought into the urban workplace to heal and calm white-collar employees.

7. Yoshimi points out that it was primarily young people, and especially single women, who were the readers or visitors of the city-as-media strategy. In a chapter interlude, I describe how a Tokyo department store aimed at middle-aged women is using sound design to appeal to this specific consumer group.

8. Amid the historic preservation movements of the 1980s, avant-garde artist Akasegawa Genpei and architecture historian Fujimori Terunobu founded the Street Observation Science Society, an association that led ethnographic and surrealist wandering expeditions throughout the city to seek out fragments of Tokyo's past (Sand 2008). As part of this association, they discovered oddities in the urban environment, first described as hyper-art (*chō-geijutsu*) that they defined as "inexplicable protuberances and concavities connected to buildings and streets in the city, which, while purposeless, have been beautifully preserved" (Sand 2008, 380): stairways leading to nowhere, for example. Akasegawa later renamed hyper-art objects "Tomason" as a tongue-in-cheek nod to Gary Thomasson, American baseball player for the Tokyo-based team Yomiuri Giants from 1981 to 1982, who reportedly played so poorly that he became nicknamed the "electric fan" for never making a hit. Tomason refers to properties of the built environment that are no longer seen as having any discernible function or purpose, yet they remain and continue to be maintained. However, is it possible to conceive of a sonic Tomason? See Yoshimura (1990).

9. See Manea (2017) for analysis of the sonic environment of Japanese trains.

10. I discuss the notion of hyperesthesia in relation to sound design practices in Tokyo in Chapter 4.

11. A sound designer employee in Tokyo told me that she is sometimes contacted by patrons of stores where they are in charge of designing and producing the store's BGM. She said she is happy when they tell her they are enjoying the store's background music.

12. My approach to BGM in Japan follows Jones and Schumacher, who describe their position vis-à-vis the bifurcated discourse regarding muzak: "In striving to avoid both of these positions, our intentions here are not to critique muzak as *music* per se, nor to reduce muzak to an unambiguous reflection of a monolithic, 'dominant ideology'" (1992, 157). As Roquet rightly points out, "Ambient media may remain limited as tools of critique, but by providing 'calm, and a space to think' they are never simply operating as a social anesthetic" (2016, 178).

13. See website for Sonic Japan: Studies in Sensory Anthropology, a digital repository for sonic life in contemporary Japan, at sonicjapan.clab.org.au.

14. One of the earliest seeds for this project came from the audio recordings of my interviews with participants in Tokyo's experimental music community during my dissertation fieldwork. Most of my interviews were conducted at cafés throughout the city, and as I later listened to and transcribed the recordings, I was struck by the constant presence of BGM in nearly all of the recordings. While it was occasionally intrusive at the time (I recall having to leave some cafés during interviews because the BGM was too loud), I later became intrigued by the atmospheric techniques behind the staging of BGM in the city.

ONE Learning to Listen to *Onkyō*

1. I conducted ethnographic fieldwork at Off Site from 2004 to 2006, during which time I regularly attended live performances and chatted informally with participants in the music scene. I also conducted formal interviews with Off Site musicians, audience members, and the owner, Itō Atsuhiro. I stayed in close contact with him after Off Site closed in 2005 and conducted follow-up interviews during summer research trips to Tokyo.

2. It is unclear the extent to which the Japanese public (specifically, residents of the major urban centers) is opposed to the ubiquitous presence of muzak. Many people I spoke with expressed annoyance and frustration at this constant sonic blanket; however, there is not an anti-muzak protest group, as seen in the UK with the emergence of Pipedown, the anti-piped-music citizen's group. See Novak (2013) for a discussion of citizens' group protests of outdoor karaoke noise in Osaka.

3. For work on the rupture of civil society in opposing the public, see Canetti (1962) [1960]; LeBon (1960) [1895]; Mazzarella (2010). For work on noise as a social threat, see Attali (1977); Russo and Warner (1987).

4. Here, we can think of the outcry over the sound of the vuvuzela at the 2010 FIFA World Cup. In this case, much of the public's antagonism (often by Westerners, perhaps unfamiliar with the sonic dimension of these trumpets) toward the sound was focused on noise pollution; the noise of the vuvuzelas was not only perceived as distracting to the spectators, but also seen as a threat to one's health on the level of noise pollution.

5. In his essay "One-Way Street," Walter Benjamin strongly recommends that writers avoid background music: "In your working conditions avoid everyday mediocrity. Semirelaxation, to a background of insipid sounds, is degrading. On the other hand, accompaniment by an etude or a cacophony of voices can become as significant for work as the perceptible silence of the night. If the latter sharpens the inner ear, the former acts as touchstone for a diction ample enough to bury even the most wayward sounds" (1978, 80).

6. Tom Gunning argues that this transformation of modern bodily experience resides in "techniques of circulation" and mobility that were directly facilitated by the development of the railway system and the advent of motion pictures (1995).

7. In contrast to Benjamin's notion of the distracted and absent-minded viewing public turned critic as one of liberation, Adorno viewed distracted listening as regressive listening. With the commodification and standardization of popular music, listening, Adorno argues, has become "arrested at the infantile stage" (cited in Leppert 2002, 303), wherein the distracted audience isn't actually listening, but is listening atomistically, as childish consumers. Thus, "regressive listeners behave like children" by focusing on fetishized (that is, atomized) elements and motifs that come to stand in for the whole (Leppert 2002, 248).

8. In his discussion of modern distraction, Jonathan Crary argues that attention and distraction are not mutually opposed; distraction is an effect of promotion and regulation

toward maintaining attention. His analysis uncovers a seemingly contradictory element of modern attention. For Crary, the notion of attention implicitly contains a sense of both tension and suspension; it evokes the sense of "holding something in wonder or contemplation." The "attentive subject is both immobile and ungrounded," yet because attention is also informed by a sense of suspension, otherwise figured as interruption or disturbance, attention contains within itself the "negation of perception itself" (1999, 10).

9. See Novak (2010) and Plourde (2008) for broader ethnographic analyses of Off Site.

10. Some regular Off Site performers I spoke with noticed a distinct change in the audience's listening habits during its five years. Off Site's history began to be reexamined during its final year of operation (2004–2005), which was rejuvenating (*moriagatta*) and provided a powerful sense of ending. This shift perhaps explains the increase in new audience members during its final year.

11. Similar to the type of experimental music discussed by Peterson, music performed at Off Site evoked a non-hierarchical aesthetic; musicians performed on the same level as the listeners, with no leader or conductor. Such forms are often referred to as non-idiomatic improvisation, in contrast to idiomatic improvisation, which is ultimately concerned with expression of a particular idiom or form such as jazz or flamenco (Bailey 1992 [1980], xi). Many of the performers at Off Site had no formal musical training (although some had visual arts backgrounds), and some preferred to refer to themselves as artists rather than musicians.

12. After Off Site closed its doors in 2005, a regular performer I interviewed used the phrase "strange and curious" (*fushigi na*) to refer to Off Site as we discussed the auditory possibilities of the space.

13. Of course, emphasizing the significance of eventfulness at Off Site would seem to implicitly acknowledge that there is a discernible connection between the performance space and the development of *onkyō* itself.

14. In addition to shutting out light, aromas were controlled. In contrast to much of Japanese public space, smoking was not permitted in the first floor performance space at Off Site. This gave the space a radically distinct sensation in comparison to the sensory dimensions of most entertainment spaces in Japan, which are sensorially defined first and foremost by veils of smoke. The lack of smoking at Off Site gave the space a sense of neutrality, an "olfactory nowhereness" (Drobnick 2005, 267), one that perhaps contradicted the emphasis on the freedom of the space and its embrace of outside environmental sounds.

15. I interviewed the owner of a classical music café, who told me that people's voices inside the café are "annoying," sonic threats to the café's calm environment.

16. His mention of "patronizing tones" refers to the Japanese command form (*~nasai*), a grammatical form that is often used by parents when scolding their children. For this listener, this form inflicts a kind of violence on the receiver, and for him is akin to disruptive, violent noise, thus, unwanted sound.

17. The studiousness of the listening public in Japan can be compared with the figure of the *otaku*, an obsessive, usually male fan and collector of manga or anime. In the case of Tokyo's experimental music scene, I rarely heard listeners referred to as *otaku*. Instead, the word "maniac" (*maniakku*) was often used to refer to listeners who are obsessive in collecting knowledge (including recordings) about musicians, or who attend live performances to a heightened degree.

18. Through conversations with regular Off Site musicians and listeners, I found that many were conscious of the scene and sound becoming stagnant, with even the owner agreeing. By the end of Off Site's operation, however, a wave of new, younger audience members—many of whom were Itō's art school students—discovered Off Site and became regular audience members. Indeed, in its final year, Off Site was visibly rejuvenated as word spread of its impending closure, thus perhaps drawing a new wave of listeners hoping to see the space before it was too late. During this time, as one informant explained, there was a palpable sense of ending, so participants, including music critics and the musicians themselves, began to reexamine Off Site's history and development.

19. It should be noted that I am not invoking *onkyō* itself as a distinctly "Japanese" genre (it was, in fact, a transnational collaborative musical form), but rather, the way in which the act of listening at Off Site often provoked discussion of what many claimed to be "Japanese" modes of listening, which refers to the idea of an aural predilection for hearing the sounds of nature, often seen as an ability that no longer retains aesthetic power or value. Such notions of nostalgia and loss (that is, a sensory mode believed to be disappearing) are, in fact, necessarily modern.

20. Comparative research concerning language use in Japan focuses on language practices and socialization, arguing that Japanese tend to be more listening-oriented, while Americans are more speaking-oriented. This pedagogy was thought to contribute to the attentive and disciplined nature of *onkyō* listeners, who had been trained from an early age to "properly sit down and listen quietly," as a regular musician at Off Site explained to me.

21. It is important to note that Tsunoda's scholarship emerged during the bubble economy era of the 1980s, a period during which nativist discourses of so-called Japaneseness (*nihonjinron*) flourished in Japanese popular media.

22. This sound artist had planned to hang wind chimes outside his residence, for example, but was scolded by his wife to take them down, as she explained that such sounds would be perceived by neighbors solely as noise and would be considered an unwelcome sonic intrusion.

23. Sophie Arkette argues that Schafer's "urban prejudice" informs his negative and static view of city sound "whereby industrial, commercial and traffic sounds are deemed sonic pollutants, and subsequently allotted to the garbage heap" (2004, 161).

24. It is also important to note the impossibility of enacting a neat division between "natural" and "man-made" sounds. For example, how do we categorize mobile phone

ringtones (a putatively "man-made" sound) that evoke the sound of birds (a "natural" sound) (Arkette 2004, 162)?

25. See Dolan (2008) for a legal and historical discussion of noise regulations in contemporary Japan.

26. Many of my musician informants were opposed to the very notion of background music and would only listen to music at home while devoting their full attention, rather than passively playing music in the background.

27. Many of the listeners and musicians I spoke with were not only conscious of such sounds, but also adapted techniques to cope with them. A number of these techniques extended to listening practices at Off Site. The oppressive nature of the urban soundscape was also often equated with the physical and mental exhaustion of navigating everyday life in Tokyo. As we discussed the rise of "healing music," for example, Itō explained that urbanites in Japan—particularly in Tokyo—lived in a constant state of exhaustion. He further posited that the cacophony of the aural environment amid an unstable economy perhaps helped to explain the rise in "healing" products, especially the genre known as "healing music." See Roquet (2008) for a discussion of the healing boom in Japan, focusing in particular on what he refers to as "ambient literature."

INTERLUDE I City Noise and the Avant-Garde oo

1. Italian Futurism occupies a contradictory space within the legacy of the historical avant-garde. Here, we can recall Walter Benjamin's critique of Futurism's aestheticization of politics, which for him marked the consummation of "art for art's sake." The Futurists' complicity with and eventual support of fascism is often denied or downplayed in academic studies, at the expense of their aesthetic activities. As one Tokyo-based music critic explained to me, Futurism's allegiances to fascism were often downplayed in Japan during the 1980s Futurist boom. Instead, Futurism was promoted and circulated primarily as an aesthetic movement.

2. The intonarumori is similar in construction to the French folk instrument the hurdy-gurdy. Constructed in the form of boxes, the intonarumori features a large metal speaker (or "diffusion horn") on one end, somewhat resembling a megaphone, which projects the sound outward. A lever on the side or top of the box serves as the mechanism for changing pitch by altering the tension of the string inside the box. By controlling the lever with the left hand and turning the handle with the right hand, the player is able to change the pitch as well as the dynamic intensity.

3. Waseda University professor and music critic Konuma Jun'ichi first learned of Italian Futurism during the bubble economy era; he was quick to clarify in an interview, however, that politically, he didn't have "sympathy" for the movement during the 1980s. For him, the Futurist boom in Japan was linked with noise and industrial music, which became

highly circulated in the 1980s. He first encountered Futurist art through the Seibu department store, which marketed itself as "patron of the avant-garde" in the bubble economy era (Sand 2006, 95). Other musicians I interviewed told me of a similar interest in Italian Futurism during the 1980s and its impact on visual art and experimental music. However, they were clear in emphasizing their disdain for the Futurists' political leanings.

4. As part of this international reassessment of Italian Futurism in the 1980s, their experiments with noise were also reexamined, and since that time, roughly thirty-five reconstructions of Russolo's intonarumori have been built throughout the world.

5. See Roquet (2016) for a discussion of Akiyama's role as curator of environmental music at Osaka Expo 1970.

6. Itō first learned of Futurism in high school after hearing a song by 1980s Japanese avant-pop band Guernica, which featured the word "intonarumori" in song lyrics. As Itō explained, he looked up the word in the dictionary and was then able to imagine and envision the noise intoners. Another informant, photographer Matsukage Hiroyuki, was deeply influenced by Italian Futurist aesthetics. He formed a performance art duo in the 1980s and named it *Complesso Plastico* (using the original Italian term rendered in Japanese), in homage to Futurist painter and sculptor Fortunato Depero and his idea of a "plastic complex."

TWO "A Place Where Time Moves Slowly"

1. White describes the ubiquity of coffeehouses in Tokyo: "In the mid 1960s, there were 27,000 coffeehouses in Tokyo. The peak was reached in 1985 when there were about 155,000 cafés in Japan. By 2008 there were about 83,000 in Tokyo. There is scarcely a block without one or more kissaten, whether for community or for solitude" (2012, 146).

2. This logic whereby the café's temporality and mood is marked as slow, relaxing, and leisurely can also be seen in the marketing of cat cafés, a recent phenomenon throughout Japan's cities, and one that has been expanding globally in recent years. See Plourde (2014).

3. It is difficult to ascertain exact numbers of Tokyo's remaining classical music cafés, but I estimate that there are approximately ten cafés still in business, and the majority lie along the Chuo subway line in western Tokyo. I am counting only those cafés that operate according to standard music café rules (specifically, no talking). Such criteria in determining whether a café can be labeled a classical music café (*meikyoku kissaten*), or just simply a café (*kissaten*), is fairly subjective. For example, I am not including the Shinjuku café L'ambre (*ranburu*) in this list. While the design exudes the postwar patina of many classical music cafés, and classical music regularly plays in the background, the café itself does not discourage talking. This is made more feasible by the relatively low volume of the background music.

4. Classical music café patrons generally tend to be middle-aged and elderly men. That

said, it is not unusual to see young female or male patrons, in their twenties and thirties. They tend to be students or office workers. I also occasionally saw couples attend music cafés. Patron demographics tend to vary slightly depending on the café. For example, some cafés seemed to have more female patrons than other cafés.

5. In order to make a request, patrons must first consult the music catalogue. The catalogues at all of the cafés where I did my fieldwork were handwritten, and the individual recordings' location on the shelf was notated in the catalogue.

6. In contrast to the process of becoming a "regular" (*jyoren*) music café patron, whose taste becomes known and catered to by the master or madam, love hotel patrons are anonymous and virtually invisible to hotel employees. Transactions are faceless and impersonal.

7. Many experimental musicians, music critics, and scholars in contemporary Japan were regular patrons at jazz cafés in the 1960s, '70s, and '80s. Musician Otomo Yoshihide famously wrote about the formative experience of attending jazz cafés in his hometown of Fukushima in the 1970s, before moving to Tokyo. For him, the essential elements of the space were "2.5X6 meters space. That and a pair of huge JBL or Altec speakers, a couple hundred jazz records and a bar counter" (cited in Novak 2013, 96–97). Conversely, other experimental musicians and record store owners I spoke with viewed jazz cafés as exclusivist and elitist institutions. Musician Keiji Haino told me of his disdain for these leisure spaces, rejecting them as a kind of "cult" institution enforcing certain ideologies. Jazz cafés were also (and to some extent still are) known to be primarily male-dominated spaces. I interviewed a middle-aged female experimental live-space owner in Tokyo who told me of the patriarchal atmosphere of jazz cafés in the 1960s, describing them as unwelcoming to female patrons.

8. Thanks to Prof. Watanabe Hiroshi for his thoughts on the historical development of the term *meikyoku*.

9. Referring to the English word "maniac," the Japanese term *maniakku* is the auditory equivalent of *otaku*, the term used to denote anime or manga fans, which has negative connotations in Japanese. Similarly, *maniakku* suggests an unhealthy obsession with (or excessive interest in) music.

10. The term *kissaten* (often shortened to *kissa*) literally translates to "tea house" in Japanese, although such places primarily serve coffee. See Takahashi (1994). See White (2012) for a rich historical and ethnographic study of Japanese coffeehouses.

11. During my interview and subsequent conversations with the madam, I was told that she wanted to avoid extensive press coverage in order to keep the café patronage small. To honor that request, I chose to use the pseudonym Marguerite, for the name of the café.

12. See Schrey (2014) for analysis of analog nostalgia.

13. The inherently corporeal nature of listening is revealed in Jonathan Sterne's history of audio technologies in which actual human ears served as auditory models for the creation of these technologies, specifically Alexander Graham Bell's ear phonautograph.

Sound reproduction technologies in the nineteenth century were referred to as "machines to hear from them," as they marked the shift in models that moved from imitating the mouth to imitating the ear (Sterne 2003, 33).

14. During one café visit, I found myself in the awkward position of dealing with allergies and needing to blow my nose, an act that is already socially unacceptable in public in Japan. I sought auditory cover under the booming classical music, and tried to be as quiet and discreet as possible. The woman sitting alone at the small table in front of me (her back was to me because all the seats face forward toward the speakers) immediately whipped her head around and shot me an angry look.

15. Public spaces with no-talking rules exist outside of the music café structure. In the western Tokyo neighborhood of Koenji is a small, two-story café in which the lower level is strictly no talking. There is a sign at the front entrance indicating the rules. Those who want to talk are asked to sit on the upper level. The lower level is designed for private activity, as most tables seat only one person, thus creating private nooks for individual patrons. The café's soundscape is a blend of almost imperceptible background music and the soft gurgling of aquariums placed throughout the room. When the server comes to take an order, the customer either whispers quietly or points to the menu and nods.

16. Hosokawa and Matsuoka describe the key distinction between radio and the phonograph in prewar Japan. Radio was part of the state apparatus and enjoyed official protection, while phonographs "received little attention from officialdom, though they obviously disliked the bourgeois hobby and its associated hedonistic consumption" (2008, 42).

17. When the Sony Walkman was first released in 1979, it initially featured two headphone jacks because private listening was not yet an acceptable mode of auditory engagement (du Gay 1997, 648). This feature was soon abandoned, as it became apparent that people were not actually listening together via the Walkman's dual-jack feature.

18. While many café patrons attend alone, it is not uncommon for patrons to attend with a friend. During fieldwork at one of my regular cafés, I noticed two middle-aged men enter and sit together silently while drinking coffee. One of the men got up to use the restroom and sat in a different seat when he returned, one that was more directly in front of the speakers. He sat with his hand resting on one cheek, listening intently for the rest of the recording.

19. Many of the regular patrons at Mignon tend to be older, and often are connected with artistic or scholarly fields, such as playwriting or *rakugo* (Japanese traditional storytelling form), or are university professors. For instance, one regular patron I was introduced to is a retired literature professor who continues to frequent jazz and classical music cafés. He has been a regular patron at Mignon for more than thirty years.

20. It is noteworthy that Mignon also serves alcohol, which is not common at most classical music cafés.

21. See Plourde (2014) for analysis of the role of the cat as intermediary in the social

space of the cat café. In this case, the affects that are produced in cat cafés serve as a node through which solitary patrons can connect with cats or other patrons. A cat café employee once joked to me that if they had a sign on the door saying that all the cats were out sick that day, their regular patrons would still attend the café.

22. In the one music café that primarily plays CDs, the elderly owner explained to me that they are physically easier for her to manage than LPs.

23. As the owner explained, this also allows Violon to charge relatively low prices for coffee, 350 yen, as opposed to most cafés that charge between 600 and 900 yen.

24. The famous jazz label Blue Note began to reissue its jazz catalogue for the Japanese market through partnerships with local record companies and distributors in the 1970s. See "Blue Note in Japan," March 2016, available at londonjazzcollector.wordpress.com.

25. I wrote in my field notebook that this recording radically transformed the atmosphere of the room and I hoped that there were other patrons in the café so I could see how they responded to this highly atypical selection.

26. During one of my fieldwork trips to Tokyo, the master at Violon invited me to a rehearsal for an upcoming gramophone concert featuring SP recordings of Argentine tango to be played before a live audience. I initially observed in the background while the master and collector played and discussed the recordings, yet the rehearsal quickly turned into a lesson on the gramophone for both the collector and me, as the master shared his extensive knowledge of vintage audio equipment and SP recordings.

27. Based on a short visit to Tokyo in 1980, ethnomusicologist Charles Keil wrote about mass culture and performance in Japan in terms of what he proposed as a cultural sensibility of "mediated and live" performance (1994).

28. This patron is a regular at several classical and jazz cafés and attends them each several times per week. He has been attending these cafés for over thirty years.

29. While occasional non-diegetic filler music is used to punctuate sensory associations or dramatic flashes of activity unfolding in the café, it is significant that there is no actual BGM playing inside the café. This is an unusual choice by the show's producers, for these types of cafés are rarely without BGM playing in the background.

30. Some of the positive English-language coverage of *Samurai Gourmet* concerning the show's slow pace and gentle qualities recalls associations of other media forms, such as *The Joy of Painting with Bob Ross*, that are believed to trigger pleasurable responses in the viewer, often explained as ASMR (autonomous sensory meridian response). For a review, see "Netflix's Latest Food Show Is Like 'Anthony Bourdain' with a Samurai Twist," March 29, 2017, available at www.thrillist.com.

31. "Pure cafés" (*junkissa*) denoted cafés that did not serve alcohol or feature erotic waitresses. In the early twentieth century, "the *kissaten* began to diverge from the café, which took on cabaret qualities"; cafés became known as "places of more gaiety and

noise than the quiet *kissaten*, where no alcohol and no erotic waitresses disturbed the contemplative setting." By contrast, the *junkissa*, or "pure *kissaten*," were the "haunts of artists, writers, and intellectuals" (White 2012, 29).

INTERLUDE II New Experiences in a New City for New Women

1. I discuss sound design techniques used in the NEWoMan restroom in greater detail in Chapter 4.

THREE "Feeling Uncomfortable Without Sound"

1. A handful of other companies provide BGM streaming services, most notably Can Systems (founded in 1966); however, USEN is the most widely known and has the biggest market share in Japan. In contrast to this, there are several high-end sound design companies based in Tokyo that focus on sonic branding, background music, and ambient sound for upscale hotels, boutique stores, bars, and restaurants. Despite the sonic ubiquity of USEN's services in retail spaces throughout Japan, there is an audible undercurrent of noncorporate, non-USEN-sponsored BGM that can still be heard in shopping arcades or older postwar-style coffeehouses (*kissaten*), among other nostalgic spaces.

2. Anahid Kassabian positions the more familiar industrial corporate history of Muzak against its counter-history, which is situated within the broad lineage of ambient and environmental music, beginning with Satie's notion of furniture music and moving through Cage and Eno (2001).

3. See Novak (2015) for a broad discussion of noise in Japan.

4. In my interviews with USEN employees, they were clear to distinguish SDO from the practice of music therapy. They do not conceive of their sound design program as medical treatments; instead, they use music to change one's hopes or aspirations (*kibou*).

5. See Roquet (2016) for a provocative analysis of ambient media as a form of mood regulation in contemporary Japan.

6. While Pandora offers the music streaming program "Pandora for Business," it is geared toward retail settings for branding and mood enhancement and is not specifically tailored to the workers themselves.

7. These mandatory stress checks apply to companies with more than fifty employees.

8. See jsca.co.jp/English for discussion of the new stress-check requirements for office workers.

9. The discourse of "work-life balance" was first promoted in 2008 by the Japanese government, specifically, the Ministry of Health, Labor, and Welfare. MHLW created Japan's work-life policy as a means of tackling Japan's social and demographic challenges,

including a rising elderly population and a falling birth rate. Japan ranks above only Hong Kong for the lowest birth rate, which is compounded by a life expectancy that is the highest in the world. With the shift to irregular workers, "less money is entering the coffers for pensions and social security" (Allison 2013, 36).

10. The name "Muzak" is Squier's neologism derived from the words "Kodak" (a booming business in the 1920s) and "music." The word became a brand name like Kleenex or Vaseline that (often disparagingly) evokes an "entire music genre" (Vanel 2013, 46).

11. The BBC's 1940 program *Music While You Work* broadcast live music to factory workers to soothe those "returning to work after nights of bombardment, thereby distracting them from dwelling on their predicament" (Sumrell and Varnelis 2007, 112–13). See Hui (2014) for a historical analysis of muzak programming in relation to US labor history.

12. In line with the shift to audio architecture, the Stimulus Progression was eventually replaced in favor of Quantum Modulation (QM), a mode of music programming that "programs foreground music to stimulate store browsers over a short period" (Anderson 2015, 823). Rather than vary throughout the workday, QM programming remains consistent in terms of musical intensity. Anderson compares QM to the Pandora algorithm, which offers an "undifferentiated horizontal line for affect management: track after track of music similar to but slightly different from the algorithm's originating seed" (2015, 823).

13. Muzak Corporation declared bankruptcy in 2009. In 2011, it was bought by the Canadian company Mood Media, which provides multimedia services to businesses, shopping malls, and dentist offices. Muzak's kitschy lowbrow image is now being rebranded under the broad rubric of sensory marketing by offering companies a range of multisensory experiences, including music, digital signage, and scent marketing. Echoing Muzak's early claims of utopian futurity, Mood Media claims on its website to provide businesses with "commercial music solutions and technology designed for today . . . and tomorrow."

14. Victorian intellectuals, part of a burgeoning middle class, raged against the street noise of organ grinders. Picker describes the anxieties and insecurities experienced by this growing (and male) class of professionals who were relegated to working from the home, which was seen as a feminized space. These writers experienced the sound of street music that filtered into their domestic work space as violently disruptive to their work. Rather than using music to enhance mood or heighten concentration, these intellectual laborers demanded silence within a "work space that sounds could not infiltrate" (Picker 2003, 429). It is important to note that many Victorian intellectuals at this time rejected the sounds of organ grinders not merely on musical grounds, but on class and ethnic grounds, for most organ grinders were of a lower class and were often immigrants.

15. See coffitivity.com. Thanks to Brendan Rooney for alerting me to this app.

16. See Roquet (2016) for a discussion of the history of BGM in postwar Japan.

17. Hostess clubs have long been crucial sites for entertaining business clients and

employees since the early postwar era in Japan. In these early leisure spaces for healing (*iyashi*), affect is managed by the eroticism of the hostess. See Allison (1994).

18. It is not surprising that the original impetus for muzak in Japan would emerge from a desire to smooth over gaps or silences and to avoid potentially awkward breaks in conversation. In 1917, composer Erik Satie envisioned furniture music (what some consider to be the precursor to muzak) as "masking the clatter of knives and forks without drowning it completely, without imposing itself. It would fill up the awkward silences that occasionally descend on guests. It would spare them the usual banalities" (cited in Lanza 2004, 17).

19. Currently, USEN features a range of channels specifically devoted to supermarkets and cafeterias.

20. I discuss how USEN is marketing BGM for the household in Chapter 4.

21. "Colorfully fruitful [*saishoku*] BGM" uses the character referencing food, which is homonymous with the Japanese word for colorful or illuminating.

22. Japan first imported French popular culture, especially music, in the early twentieth century. Parisian music became popular in postwar Japan as many *chanson* (French popular song) cafés opened, though very few still remain. Despite this, French culture is still nostalgically embraced in contemporary Japanese consumer culture, as can be seen by the numerous French bakeries, crêperies, and cafés throughout Japan, especially Tokyo. It is perhaps not surprising, then, that this BGM channel would explicitly evoke the sounds and imagined ambience of France to relax office workers.

23. Though only a handful of high schools subscribe to SDO, the impetus is similar to that of the corporate world: to reduce stress for students and increase efficiency. BGM is played in the cafeteria, in the mental health counseling center, and through speakers mounted in the hallway. The USEN manager told me that since subscribing to SDO, teachers no longer have to yell at students to stop running in the hallways, a common occurrence before the introduction of background music.

24. See the short 2014 documentary by Shiho Fukada, *Overworked to Suicide*. Available at mediastorm.com.

25. Most new employees start work in April, which is also when the new school year begins. "May sickness" is believed to occur after the end of the national holiday "Golden Week" in early May.

26. Of course, office architecture in North America has been shifting toward a more open-air layout, especially prominent in the tech industry. The difference here is that workers can use headphones to block out co-workers and create their own private moodscapes.

27. Citing Gilles Deleuze, Sumrell and Varnelis track the shift in office architecture as an effect of the transition from a society of discipline to a society of control. Muzak played an important role in this shift. During the wartime era, as theories of surveillance were predominant, muzak served to discipline and oversee workers while symbolically

reminding them that "the boss was thinking of them" (2007, 111). In the postwar era, cybernetic theories provide the impetus for a new logic of management. Instead of being under surveillance, workers were now given more responsibility. This discursive shift demanded a new open office layout: "In order for employees to share information and expertise, social interaction became crucial and open environments replaced closed offices and executive floors" (2007, 120). Of course, muzak then took on a new role within these open office plans; it was now tasked with masking distracting background sounds that were circulating freely throughout offices.

28. Sonic masking has been used in other contexts in Japan, most notably in the "sound princess" (*oto hime*), a white-noise device installed in women's restroom stalls to cover up sounds of personal hygiene.

29. I interviewed a worker whose company subscribes to SDO, and she (somewhat laughingly) told me that her office occasionally uses this channel, but primarily to celebrate a co-worker's birthday.

30. This same doctor authored a study in a USEN brochure on music for concentration, where he advocates the "Mozart effect," a popular phenomenon during the 1990s, where it was believed that one's spatial reasoning would strengthen as a result of listening to Mozart. This study was conducted in 1993 and later discredited and widely debunked.

31. For a full tally of channels in existence at the time of research, refreshment has twenty-two, relaxation has twenty-three, awareness has twenty, and concentration has eleven channels.

32. The one channel in the relaxation category that might initially seem at odds with the therapeutic dimension of the majority of the channels is the "Salon Classic" channel, the concept of which is a "salon for adults" featuring refined and lyrical piano music. Most of the composers featured are Romantic or late Romantic era, including a clever nod to the sensibilities of composer Erik Satie in the channel description. Whether SDO listeners are aware of Satie's connection to the larger genealogy of ambient music is unclear, yet his inclusion seems significant.

33. An employee I interviewed whose company uses SDO told me that the music featured in the relaxation category was so relaxing for the workers that it became difficult to concentrate on work.

34. The term "easy listening" in Japanese does not have the negative connotations it does in the West, where it is often mocked as lowbrow and kitschy. Instead, as USEN uses the term, it denotes exactly what the name itself evokes: soft, gentle music that is easy to listen to. Less a narrow genre label, easy listening refers more broadly to stylistic musical traits. Therefore, classical music or contemporary pop music could be considered easy listening. That sense of ease is crucial to the efficacy of SDO and its creation of an affective atmosphere.

35. A worker whose office subscribes to SDO told me that the company has recently

discussed the possibility of canceling the service as a cost-cutting measure. She told me that she is completely opposed to this: "I love the BGM!" she laughingly admitted.

36. *Kūkan* is a more abstract way to express filling a space or void.

37. Roquet describes how the Japanese term for atmosphere (*funiki*) has undergone a transformation similar to that of the English word, expanding in the Meiji period (1868–1912) to "refer not just to a pocket of air but to the subjectively felt feeling and tone of a place" (2016, 3). This expansion in meaning was an attempt to "keep pace with the European notion of the feeling or mood of a place" (2016, 187).

FOUR Sonic Air Conditioning

1. See Anderson (2015) and Lanza (2004) for use of the phrase "sonic air conditioning." I use the phrase to evoke the infrastructural and highly engineered presence of background music.

2. Interestingly, I encountered the use of Pachelbel's *Canon in D* as BGM in other areas of Tokyo, although it was employed to very different ends. For example, the piece was played as closing time music at a classical music café in Tokyo that I frequented. Drawing on the extensive practice in Japan of using music (or generic sonic melodies) to sonically indicate that stores are about to open or close, this particular café plays *Canon in D* at exactly 5:55 p.m. every night, five minutes before closing. Soon after the piece begins, most café patrons begin slowly packing up their belongings to leave. At the same time, it would be extremely unusual to hear Pachelbel's *Canon in D* played in a classical music café as part of the musical offerings for the listener-patrons during business hours. Its value thus lies in its functionality rather than its role as a representative work from the Baroque era to be listened to attentively by café patrons. Here, the music's function is to aurally discipline patrons to leave the café, yet in a sonically gentle, non-forceful way.

3. See Sterne's analysis of the weaponized use of classical music used by some businesses as a deterrent for undesirable customers, a strategy he refers to as the "non-aggressive music deterrent" (2013).

4. It has drawn the ire of musicians as disparate as Ted Nugent, Genesis P-Orridge, and Brian Eno. These musicians have rejected Muzak in distinct, though occasionally overlapping ways. Nugent was strenuously opposed to it on aesthetic grounds, decrying it as "sucking the lifeblood from countless melodies since its founding." In 1989, he attempted in vain to buy the company in order to destroy it. In 1984, P-Orridge published an essay advocating the creation of an "anti-muzak" "that, instead of cushioning the sounds of a factory environment, made use of those very sounds to create rhythmic patterns and structures that incorporated the liberating effects of music by unexpected means. This approach is diametrically opposed to the position of official MUZAK." Composer and producer Brian Eno famously defined his notion of "ambient music" as an explicit

rejection of "canned music," which he viewed as "stripping away all sense of doubt and uncertainty (and thus all genuine interest) from the music."

5. Vanel describes how "criticizing Muzak for its efficiency is as absurd as blaming teenagers' suicide on the influence of heavy metal music—as did the Parent Teacher Association in the 1990s—while ignoring that the source of the problem lies elsewhere" (2013, 4). He argues that the company's own marketing of its product as increasing productivity has "been accepted uncritically by its harshest opponents" (2013, 5).

6. This is contrasted with USEN's Sound Design for Office program, discussed in the previous chapter. Much of the corporate marketing of the program draws on pseudoscientific testing to support claims that this music program reduces stress for workers and makes them more productive.

7. While air conditioning is not necessarily condemned in the ways that other ambient technologies like muzak are, it is acknowledged as having a deep impact on cultural practices. See Arsenault (1984). Thanks to Gretchen Bakke for informing me of this reference.

8. Marks & Spencer announced in June 2016 that it will adopt a music-free policy across three hundred of its stores in the UK following complaints from customers and employees about the in-store sonic repetition. See "M&S to Turn Off Music in Stores Following Customer Feedback," June 1, 2016, available at www.theguardian.com/business. Thanks to Ethan Spielman for alerting me to this article.

9. Many companies that subscribe to USEN's Sound Design for Office program were reluctant to talk to me about their experiences. This reluctance seemed indicative of the general difficulty of talking about their relationship to sound and its effect on their labor practices. Given USEN's strong corporate brand, I also wondered if some companies that may have had negative experiences with SDO (or canceled the service) were reluctant to share those opinions publicly.

10. It is perhaps not surprising that many of these employees expressed interest in reading my analysis of BGM listening practices in Tokyo, perhaps as a means of advertising for the company. For example, I conducted an interview with a high-end sound design company, and prior to the interview, I was asked if employees could interview me about my research on BGM in Japan. This interview was later featured on the company blog.

11. It is not surprising that certain sensual properties of commodities have become privatized and trademarked. Harley Davidson, for example, has sought to trademark the infamous sound of its motorcycle engines, and Singapore Airlines has patented its "warm towel smell" (Howes 2005, 288; Malefyt 2009, 17). See also Powers (2010).

12. Curiously, sushi restaurants (and sometimes soba) do not typically have BGM. Most of the sound designers I spoke with believed that the "natural" sounds of food being prepared in these types of restaurants serve to create a unique "musical" soundscape, thus rendering BGM unnecessary. Ramen spots also do not typically have BGM, but I was told that this is because ramen is seen as food to be eaten quickly and not lingered over, as

would occur at other more formal Japanese-style restaurants; therefore, BGM serves little purpose in targeting individual patrons. Despite this argument about BGM as antithetical to businesses built on expediency, such as ramen, nearly all fast-food chains in Japan have BGM, often sponsored by USEN.

13. Here, the designer is referring to the legendary *enka* singer and Japanese cult actress Kaji Meiko, whose music was prominently featured in both *Kill Bill* soundtracks.

14. Hotels are the classic example of what anthropologist Marc Augé referred to as "non-places," that is, liminal and transitory public spaces such as airport waiting rooms, fast-food restaurants, and supermarket checkout lines that people pass through rather than inhabit. Antithetical to the notion of a place, non-places "cannot be defined as relational, or historical, or concerned with identity" (Augé 1995, 77–78). While places are "organically social" and rooted in historical significance, non-places are necessarily grounded in a kind of "solitary contractuality" (1995, 94). It is not surprising, then, that chain hotels and airports are classic examples of non-places for Augé, "being strangely familiar in every corner of the globe and instinctively comprehensible to all supermodern subjects, regardless of language or other markers of local identity" (Hainge 2009, 203). Given this, one might expect a uniformity of sound in these non-places, as if all hotels would sound the same.

15. YouTube features numerous videos of *shishi-odoshi* as a form of white noise or tinnitus relief. One video features a "three-hour Japanese fountain 'Shishi-odoshi' sound for relaxation and healing."

16. See Novak (2010) for a discussion of the complex reappraisal of *ma* for postwar avant-garde composers in Japan.

17. Bossa nova has long been used as elevator music, and it has also served as musical inspiration for musicians like Burt Bacharach, whose music forms the core of postwar elevator music. See Scannell (2014).

18. Jazz occupies a complex position within Japan's musical landscape—jazz *kissaten* (cafés) are well known for their dedicated listeners, and the domestic recording industry serves as a global haven for jazz record collectors. Despite this, jazz is one of the most common forms of BGM in leisure spots such as bars and restaurants.

19. See Yano (2002) for a rich ethnographic analysis of *enka.*

20. Toilets have long been asserted as emblematic of Japan's aesthetic and, more recently, its technological advancement. Writer Tanizaki Jun'ichiro's 1933 essay on Japanese notions of beauty, "In Praise of Shadows," features a section somewhat parodically reflecting on the traditional Japanese toilet as a meditative and tranquil place of "great spiritual repose." He contrasts this with the garishly lit Western toilet. The current global popularity of toilets by Toto suggests the recognition and consumer interest in technologically advanced toilets offering such features as heated seats, bidets, and various sound options.

21. The department store name is a portmanteau of "New" and "Woman," pronounced

as *nyūman* in Japanese. The store's name announces its consumer focus on women; in this case, women in their forties and beyond. NEWoMan distinguishes itself from other department store chains geared toward younger women in their twenties and thirties, such as Lumine and Marui.

22. In a sense, this evokes the practices of other subcultural groups such as the train-fan (*tecchan*) subculture, a subset of which is obsessed with the sound of trains and is devoted to making live field recordings (*nama roku*) of trains.

23. For recent work on this emergent fantasy of sociality in post-Fukushima Japan, see Nozawa (2015).

24. Pandora, Spotify, Moodagent, Songza, iHeartRadio, Beats, Musicovery, and Bipolar Radio, among other streaming services, all offer mood-based playlists (Anderson 2015, 828).

25. In an online promotional magazine, BGM company Monstar Channel describes the ways in which different genres of jazz can be engineered in distinct consumer spaces, including ramen spots and teppanyaki. The article concedes that it might seem unexpected for a ramen or teppanyaki restaurant to play jazz, but in fact, "modern jazz" fits perfectly (*pittari*) and will help the business evolve. The company recommends playing contemporary jazz or jazz hip-hop, which will be "easy to listen to," especially for patrons who don't listen to music very much.

26. Paul Roquet examines Muji's marketing of "no-style style" as indicative of the "subtractivist" style that was popular in the early 1980s as a backlash to the hyper-consumption of the bubble economy era (2016, 136).

27. See "Haruomi Hosono—Watering a Flower," June 6, 2012, available at glob.daniel-letson.com. Thanks to Peter Kaiser for alerting me to this article.

28. Muji has also released a relaxation app. The Muji to Relax app offers users the choice of six nature soundscapes to choose from: waves crashing, birds chirping, crackling fire, mountain stream, forest sounds, waterfall. The app also offers users the option to measure heart rate (by placing one's finger over the smartphone's camera lens), after which the app will "create your own relaxing music with your pulse and the time."

29. The record label owner told me that they had received complaints from Tower Records Japan for putting the half-off label on the CD cover, because this could be potentially confusing for record store patrons.

30. It is intriguing that baby's breath was chosen as the flower in this image. Long considered to be a "filler" generic flower that serves as backdrop in most bouquets, baby's breath is never used as the centerpiece. It is perhaps the floral equivalent of BGM.

31. See Carpenter (2013) for analysis of the use of muzak in *Dawn of the Dead*.

32. To hear a track from *Music for Supermarket*, visit soundcloud.com and search "Music for Supermarket Sweet Dreams Press."

33. More recently, there has been a global undercurrent of fascination with this kind of lo-fi retro ubiquitous music, as can be seen with the vaporwave phenomenon. Relatedly, in 2016, Mark Davis, a former employee of Kmart who worked there from 1989 to 1999, uploaded fifty-six hours of digitized cassettes of in-store music and advertising. The majority of the user comments suggest longing and nostalgia for the loss of this in-store music programming. Such expressions of nostalgia are also bound up with Kmart's economic downfall; the company declared bankruptcy in recent years. One user wrote, "I can't really explain but I've always had this strange fascination with old elevator music. Perhaps the way it takes you back to the good ol days when things were a lot simpler." Another user vividly described the childhood sensorial associations of the store itself: "Kmart at Christmas time was one of my fondest retail memories as a kid. Those aqua colored uniforms and the vintage cafeteria with specials like hot turkey dinners. The pumpkin pie was mandatory. And this cheesy music wafting through the deliciously fragrant air of hot processed holiday food and brand new merchandise. It's so depressing shopping at Kmart now. It's always empty, the shelves are a mess, the music is 5 Seconds of Summer, the Little Caesar's pizza stinks and they're undersold by Walmart." See archive.org and search "Kmart music."

CONCLUSION Tokyo Listening, Listening to Tokyo

1. See Abe (2018) for an ethnographic analysis of the economic impact of *jishuku* on *chindon-ya* practitioners whose livelihood depends on the ability to make noise in public.

2. See Giaimo (2016) for discussion of Irv Teibel, the central figure behind the creation of environmental recordings in the 1970s as a means of self-soothing, relaxation, and increased productivity.

3. The Japanese beef bowl chain, Tokyo Chikara Meshi (similar to Yoshinoya), has been dealing with recent economic losses and subsequently has closed a number of locations nationwide, including many in Tokyo. After joking with an acquaintance that perhaps this is because they don't always play BGM in their stores, I was surprised to read negative reviews on Tabelog and Twitter that specifically mention the store's lack of BGM as creating an uncomfortable environment. One reviewer bemoaned the lack of BGM as forcing people to engage with the undesirable sounds of eating. As the reviewer wrote, "Even though it's a chain store, there was no BGM playing, and there was only the clanking sounds of eating utensils which created a feeling of disorder or malfunction." Another reviewer stated, "When I first entered, I was really surprised at this completely silent space since there was no BGM playing." Both comments reveal how BGM is an expected sonic backdrop that evokes normalcy in consumer spaces (especially chain stores); thus, when there is no BGM, there is a feeling of something "going wrong." Thanks to Shunsuke Nozawa for pointing out the lack of BGM in Tokyo Chikara Meshi.

BIBLIOGRAPHY

Abe, Marié. 2014. "Sounding Imaginative Empathy: *Chindon-ya*'s Affective Economies on the Streets of Osaka." In *Sound, Space, and Sociality in Modern Japan*, edited by Joseph D. Hankins and Carolyn S. Stevens, 89–107. London: Routledge.

———. 2018. *Resonances of Chindon-ya: Sounding Space and Sociality in Contemporary Japan*. Middletown: Wesleyan University Press.

Adey, Peter, Laure Brayer, Damien Masson, Patrick Murphy, Paul Simpson, and Nicolas Tixier. 2013. "'Pour votre Tranquillite': Ambiance, Atmosphere, and Surveillance." *Geoforum* 49: 299–309.

Adorno, Theodor W. 2002. *Essays on Music*. Edited by Richard Leppert. Translated by Susan H. Gillespie. Berkeley: University of California Press.

Akasegawa, Genpei. 2009. *Hyperart: Thomasson*. Translated by Matthew Fargo. New York: Kaya.

Allison, Anne. 1994. *Nightwork: Sexuality, Pleasure, and Corporate Masculinity in a Tokyo Hostess Club*. Chicago: University of Chicago Press.

———. 2013. *Precarious Japan*. Durham: Duke University Press.

Anderson, Ben. 2009. "Affective Atmospheres." *Emotion, Space and Society* 2: 77–81.

Anderson, Paul Allen. 2015. "Neo-Muzak and the Business of Mood." *Critical Inquiry* 41: 811–40.

Apollonio, Umbro, ed. 1970. *Futurist Manifestos*. Translated by Robert Brain, R. W. Flint, J. C. Higgitt, and Caroline Tisdall. Boston: MFA Publications.

Arkette, Sophie. 2004. "Sounds Like City." *Theory, Culture and Society* 21(1): 159–68.

Arsenault, Raymond. 1984. "The End of the Long Hot Summer: The Air Conditioner and Southern Culture." *Journal of Southern History* 50(4): 597–628.

Atkins, E. Taylor. 2001. *Blue Nippon: Authenticating Jazz in Japan*. Durham: Duke University Press.

Attali, Jacques. 1977. *Noise: The Political Economy of Music*. Translated by Brian Massumi. Minneapolis: University of Minnesota Press.

Augé, Marc. 1995. *Non-places: Introduction to an Anthropology of Supermodernity*. London: Verso.

Bailey, Derek. 1992 [1980]. *Improvisation: Its Nature and Practice in Music*. Cambridge, MA: Da Capo.

Beer, David. 2007. "Tune Out: Music, Soundscapes and the Urban *Mise-en-Scène*." *Information, Communication, and Society* 10(6): 846–66.

Bell, Clive. 2003. "Site for Sore Ears." *The Wire* July 2003: 38–45.

Belson, Ken, and Norimitsu Onishi. "In Deference to Crisis, a New Obsession Sweeps Japan: Self-Restraint." *New York Times*, March 27, 2011. Available at nytimes.com.

Benjamin, Walter. 1968. *Illuminations: Essays and Reflections*. Edited by Hannah Arendt. Translated by Harry Zohn. New York: Schocken.

———. 1978. *Reflections: Essays, Aphorisms, Autobiographical Writings*. Translated by E. Jephcott. New York: Schocken.

Bickford, Tyler. 2014. "Earbuds are Good for Sharing: Children's Headphones as Social Media at a Vermont School." In *The Oxford Handbook of Mobile Music Studies*, edited by Jason Stanyek and Sumanth Gopinath, 335–55. Oxford: Oxford University Press.

Bijsterveld, Karin. 2008. *Mechanical Sound: Technology, Culture, and Public Problems of Noise in the Twentieth Century*. Cambridge, MA: MIT Press.

———. 2010. "Acoustic Cocooning: How the Car Became a Place to Unwind." *Senses and Society* 5(2): 189–211.

Bissell, David. 2008. "Comfortable Bodies: Sedentary Affects." *Environment and Planning A* 40(7): 1697–1712.

Böhme, Gernot. 1993. "Atmosphere as the Fundamental Concept of a New Aesthetics." *Thesis Eleven* 36: 113–26.

———. 2017. *The Aesthetics of Atmospheres*. Edited by Jean-Paul Thibaud. London: Routledge (e-book).

Boutin, Aimée. 2015. *City of Noise: Sound and Nineteenth-Century Paris*. Urbana: University of Illinois Press.

Bull, Michael. 2007. *Sound Moves: Ipod Culture and Urban Experience*. London: Routledge.

Bull, Michael, and Les Back, eds. 2003. *The Auditory Culture Reader*. New York: Berg.

Cage, John. 1961. *Silence: Lectures and Writings*. Cambridge, MA: MIT Press.

Canetti, Elias. 1962 [1960]. *Crowds and Power*. New York: Viking.

Carpenter, Alexander. 2013. "Dead in Tune: Uncanny Muzak in *Dawn of the Dead*." *Journal of Popular Culture* 46(6): 1231–52.

Clancy, Patricia. 1986. "The Acquisition of Communicative Style in Japan." In *Language Socialization across Cultures*, edited by Bambi Schieffelin and Eleanor Ochs, 213–50. Cambridge: Cambridge University Press.

Connor, Steven. 2004. "Edison's Teeth: Touching Hearing." In *Hearing Cultures: Essays on Sound, Listening, and Modernity*, edited by Veit Erlmann, 153–72. Oxford: Berg.

Cook, Haruko Minegishi. 1999. "Language Socialization in Japanese Elementary Schools: Attentive Listening and Reaction Turns." *Journal of Pragmatics* 31: 1443–65.

Corbin, Alain. 1998. *Village Bells: Sound and Meaning in the 19th Century French Countryside*. New York: Columbia University Press.

Coulthard, Lisa. 2012. "Haptic Aurality: Listening to the Films of Michael Haneke." *Film-Philosophy* 16(1): 16–29.

Cox, Christoph, and Daniel Warner, eds. 2004. *Audio Culture: Readings in Modern Music*. New York: Continuum.

Crary, Jonathan. 1999. *Suspensions of Perception: Attention, Spectacle, and Modern Culture*. Cambridge, MA: MIT Press.

Deleuze, Gilles, and Felix Guattari. 1987. *A Thousand Plateaus: Capitalism and Schizophrenia*. Translated by Brian Massumi. Minneapolis: University of Minnesota Press.

Derschmidt, Eckhart. 1998. "The Disappearance of the *Jazu-Kissa*: Some Considerations about Japanese 'Jazz-Cafes' and Jazz-Listeners." In *The Culture of Japan as Seen through Its Leisure*, edited by Sepp Linhart and Sabine Früstück, 303–15. Albany: State University of New York Press.

Diaconu, Madalina. 2006. "Patina—Atmosphere—Aroma: Towards an Aesthetics of Fine Differences." In *Logos of Phenomenology and Phenomenology of the Logos. Book Five: The Creative Logos. Aesthetic Ciphering in Fine Arts, Literature and Aesthetics*, edited by A. T. Tyminiecka, 131–48. Dordrecht: Springer.

Diaconu, Madalina, E. Heuberger, R. Mateus-Berr, and L. Marcel Vosicky, eds. 2011. *Senses and the City: An Interdisciplinary Approach to Urban Sensescapes*. Vienna: Verlag.

Dolan, Daniel. 2008. "Cultural Noise: Amplified Sound, Freedom of Expression and Privacy Rights in Japan." *International Journal of Communication* 2: 662–90.

Drobnick, Jim. 2005. "Volatile Effects: Olfactory Dimensions of Art and Architecture." In *Empire of the Senses: The Sensual Culture Reader*, edited by David Howes, 265–80. Oxford: Berg.

Eno, Brian. 2004. "Ambient Music." In *Audio Culture: Readings in Modern Music*, edited by Christoph Cox and Daniel Warner, 94–47. New York: Continuum.

Erlmann, Veit, ed. 2004. *Hearing Cultures: Essays on Sound, Listening and Modernity*. Oxford: Berg.

Faison, Elyssa. 2007. *Managing Women: Disciplining Labor in Modern Japan*. Berkeley: University of California Press.

Foucault, Michel. 1995 [1975]. *Discipline and Punish: The Birth of the Prison*. New York: Vintage.

Foy, George Michelson. 2010. *Zero Decibels: The Quest for Absolute Silence*. New York: Scribner.

Galliano, Luciana. 2002. *Yôgaku: Japanese Music in the Twentieth Century*. Translated by Martin Mayes. Lanham, MD: Scarecrow.

du Gay, Paul, Stuart Hall, Linda Janes, Hugh Mackay, and Keith Negus, eds. 1997. *Doing Cultural Studies: The Story of the Sony Walkman*. Thousand Oaks, CA: Sage.

Giaimo, Cara. 2016. "The Man Who Recorded, Tamed, and Then Sold Nature Sounds to America." *Atlas Obscura* April 5, 2016. Available at www.atlasobscura.com.

Gitelman, Lisa. 2003. "How Users Define New Media: A History of the Amusement Phonograph." In *Rethinking Media Change: The Aesthetics of Transition*, edited by David Thorburn and Henry Jenkins, 61–80. Cambridge, MA: MIT Press.

Gregg, Melissa. 2010. "On Friday Night Drinks: Workplace Affects in the Age of the Cubicle." In *The Affect Theory Reader*, edited by Melissa Gregg and Gregory J. Seigworth, 250–68. Durham: Duke University Press.

Gunning, Tom. 1995. "Tracing the Individual Body: Photography, Detectives, and Early Cinema." In *Cinema and the Invention of Modern Life*, edited by Leo Charney and Vanessa R. Schwartz, 15–45. Berkeley: University of California Press.

Haden-Guest, Anthony. 1973. *The Paradise Program: Travels through Muzak, Hilton, Coca-Cola, Texaco, Walt Disney and other World Empires*. New York: William Morrow.

Hagood, Mack. 2011. "Quiet Comfort: Noise, Otherness, and the Mobile Production of Personal Space." *American Quarterly* 63(3): 573–89.

Hainge, Greg. 2009. "Airport Music: Muzak, 'Non-lieux' and film sound in Stéphanik's *Stand-by* and Lioret's Tombés du ciel." *Studies in French Cinema* 9(3): 201–14.

Hankins, Joseph D., and Carolyn S. Stevens, eds. 2014. *Sound, Space and Sociality in Modern Japan*. London: Routledge.

Hardt, Michael. 1999. "Affective Labor." *boundary 2* 26(2): 89–100.

Hardt, Michael, and Antonio Negri. 2004. *Multitude: War and Democracy in the Age of Empire*. New York: Penguin.

Harvey, David. 1989. *The Condition of Postmodernity*. London: Basil Blackwell.

Hayashi, Sharon, and Anne McKnight. 2005. "Good-bye Kitty, Hello War: The Tactics of Spectacle and New Youth Movements in Urban Japan." *Positions* 13(1): 87–113.

Healy, Stephen. 2014. "Atmospheres of Consumption: Shopping as *Involuntary Vulnerability*." *Emotion, Space and Society* 10: 35–43.

Hirato, Renkichi. 2005. "Manifesto of the Japanese Futurist Movement." Translated by Miryam Sas. In *Histories of the Future*, edited by Daniel Rosenberg and Susan Harding, 225–30. Durham: Duke University Press.

Hirschkind, Charles. 2006. *The Ethical Soundscape: Cassette Sermons and Islamic Counterpublics*. New York: Columbia University Press.

Holloway, Julian, and Sheila Hones. 2007. "Muji, Materiality, and Mundane Geographies." *Environment and Planning A* 39(3): 555–69.

Hosokawa, Shuhei. 2014. "The Swinging Phonograph in a Hot Teahouse: Sound Technology and the Emergence of the Jazz Community in Prewar Japan." In *Sound, Space, and Sociality in Modern Japan*, edited by Joseph D. Hankins and Carolyn S. Stevens, 108–26. London: Routledge.

———. (N.d.) "Music, Sound, and the Sonic in Japanese History." Unpublished manuscript.

Hosokawa, Shuhei, and Hideaki Matsuoka. 2004. "Vinyl Record Collecting as Material Practice: The Japanese Case." In *Fanning the Flames: Fans and Consumer Culture in Contemporary Japan*, edited by William W. Kelly, 151–67. Albany: SUNY Press.

———. 2008. "On the Fetish Character of Sound and the Progression of Technology: Theorizing Japanese Audiophiles." In *Sonic Synergies: Music, Technology, Community, Identity*, edited by Gerry Bloustien, Margaret Peters, and Susan Luckman, 39–50. Aldershot, Hampshire: Ashgate.

Hosoma, Hiromichi. 2005. "*Watashi ga oto wo kîteiru basho*" [The Place Where I Listen to Sounds]. *diatxt* (Diatext) 16. Kyoto: Kyoto Art Center.

Howes, David. 2005. "Hyperesthesia, or, The Sensual Logic of Late Capitalism." In *Empire of the Senses: The Sensual Culture Reader*, edited by David Howes, 281–303. Oxford: Berg.

Hui, Alexandra. "Muzak-While-You-Work: Programming Music for Industry, 1919–1948." *Historiche Anthropologie* 22: 364–83.

Ihde, Don. 2007. *Listening and Voice: Phenomenologies of Sound*. Albany: SUNY Press.

Inoue, Miyako. 2006. *Vicarious Language: Gender and Linguistic Modernity in Japan*. Berkeley: University of California Press.

Itō, Yukari. 2002. Gallery Lineups: [off side (*hansoku*)] na [site (*ichi*)] kara. *Bijutsu Techō* 54(821): 88.

Ivy, Marilyn. 1993. "Formations of Mass Culture." In *Postwar Japan as History*, edited by Andrew Gordon, 239–58. Berkeley: University of California Press.

———. 1996. *Discourses of the Vanishing: Modernity, Phantasm, Japan*. Chicago: University of Chicago Press.

———. 1996. "The Mystery Man with 21 Faces." *Critical Inquiry* 23(1): 11–36.

Johnson, James H. 1995. *Listening in Paris: A Cultural History*. Berkeley: University of California Press.

Jones, Simon C., and Thomas G. Schumacher. 1992. "Muzak: On Functional Music and Power." *Critical Studies in Mass Communication* 9: 156–69.

Kane, Brian. 2014. *Sound Unseen: Acousmatic Sound in Theory and Practice*. Oxford: Oxford University Press.

Kassabian, Anahid. 2001. "Ubiquitous Listening and Networked Subjectivity." ECHO*: A Music-Centered Journal* 3(2): 1–13.

———. 2013. *Ubiquitous Listening: Affect, Attention, and Distributed Subjectivity*. Berkeley: University of California Press.

Katz, Mark. 2004. *Capturing Sound: How Technology Has Changed Music*. Berkeley: University of California Press.

Kawaguchi, Yoko. 2015. *Tokyo no kissaten: Go haku iro no shizuku 77 teki* [Tokyo's Cafes: 77 Amber Drops]. Tokyo: Jitsugyo no Nihon Sha.

Keil, Charles. 1994. "Music Live and Mediated in Japan." In *Music Grooves*, edited by Charles Keil and Steven Feld, 247–56. Chicago: University of Chicago Press.

Kelman, Ari Y. 2010. "Rethinking the Soundscape: A Critical Genealogy of a Key Term in Sound Studies." *Senses and Society* 5(2): 212–34.

Kheshti, Roshanak. 2011. "Touching Listening: The Aural Imaginary in the World Music Culture Industry." *American Quarterly* 63: 711–31.

LaBelle, Brandon. 2010. *Acoustic Territories: Sound Culture and Everyday Life*. New York: Continuum.

Lanza, Joseph. 2004. *Elevator Music: A Surreal History of Muzak®, Easy-Listening, and Other Moodsong®*. Ann Arbor: University of Michigan Press.

———. 2013. "Foreground Flatland." In *The Oxford Handbook of New Audiovisual Aesthetics*, edited by John Richardson, Claudia Gorbman, and Carol Vernallis, 622–27. London: Oxford University Press.

LeBon, Gustave. 1960 [1895]. *The Crowd: A Study of the Popular Mind*. New York: Viking.

Leppert, Richard, ed. 2002. *Essays on Music* by Theodor W. Adorno. Translated by Susan H. Gillespie. Berkeley: University of California Press.

Levin, David Michael, ed. 1993. *Modernity and the Hegemony of Vision*. Berkeley: University of California Press.

Lukacs, Gabriella. 2010. *Scripted Affects, Branded Selves: Television, Subjectivity, and Capitalism in 1990s Japan*. Durham: Duke University Press.

Malefyt, Timothy de Waal. 2009. "Evaluating Sensory Experience: Self-Reporting Consumers and Commercial Fieldwork." *Anthropology News* April 2009: 17.

Manabe, Noriko. 2015. *The Revolution Will Not Be Televised: Protest Music after Fukushima*. Oxford: Oxford University Press.

Manea, Pierre. 2017. "Melodic Refrains in Japanese Train Stations: The Management of Passenger Behavior through the Use of Electric Bells." In *Toward an Anthropology of Ambient Sound*, edited by Christine Guillebaud, 98–114. London: Routledge.

Marinetti, Filippo. 1971. *Marinetti: Selected Writings*. Edited by R. W. Flint. Translated by R. W. Flint and Arthur A. Coppotelli. New York: Farrar, Straus, and Giroux.

Marks, Laura U. 2000. *The Skin of the Film: Intercultural Cinema, Embodiment, and the Senses*. Durham: Duke University Press.

Massumi, Brian. 2002. *Parables for the Virtual: Movement, Affect, Sensation*. Durham: Duke University Press.

Mazzarella, William. 2010. "The Myth of the Multitude, or Who's Afraid of the Crowd?" *Critical Inquiry* 36: 697–727.

Molasky, Michael. 2005. *Sengo nihon no jazu bunka: Eiga, bungaku, sabukaruchaa* [The Jazz Culture of Postwar Japan: Film, Literature, Subculture]. Tokyo: Seidosha.

Moles, Abraham. 1968. *Information Theory and Esthetic Perception*. Translated by Joel E. Cohen. Urbana: University of Illinois Press.

Nakajima, Yoshimichi. 1996. *Urusai nihon no watashi* [My Noisy Japan]. Tokyo: Shinchō-bunko.

North, Scott. 2014. "Limited Regular Employment and the Reform of Japan's Division of Labor. *Asia-Pacific Journal: Japan Focus*, 12(15) no. 1, April 14, 2014.

Novak, David. 2008. "2.5 x 6 Metres of Space: Japanese Music Coffeehouses and Experimental Practices of Listening." *Popular Music* 27(1): 15–34.

———. 2010. "Playing Off Site: The Untranslation of Onkyō." *Asian Music* 41(1): 36–59.

———. 2013. *Japanoise: Music at the Edge of Circulation*. Durham: Duke University Press.

Novak, David, and Matt Sakakeeny, eds. 2015. *Keywords in Sound*. Durham: Duke University Press.

Nozawa, Shunsuke. 2015. "Phatic Traces: Sociality in Contemporary Japan." *Anthropological Quarterly* 88(2): 373–400.

O'Doherty, Brian. 1986. *Inside the White Cube: The Ideology of the Gallery Space*. Berkeley: University of California Press.

Omuka, Toshiharu. 2000. "Futurism in Japan, 1909–1920." In *International Futurism in Arts and Literature*, edited by Gunter Berghaus, 244–70. Berlin: Walter de Gruyter.

Otomo, Yoshihide. 2001. *Sokkyō to Onkyō wo meguru 10 mai* [10 Selections of Improvisation and *Onkyō*]. *InterCommunication* 35. Tokyo: NTT.

———. 2005. "Filament at Off Site." April 30, 2005. Available at d.hatena.ne.jp (accessed December 26, 2006).

Peterson, Marina. 2007. "Improvising in Beirut: Nation-State Making and Cultural Diplomacy in a Post 9–11 World." Conference presentation. International Association for Popular Music Studies (IASPM).

———. 2010. *Sound, Space, and the City: Civic Performance in Downtown Los Angeles*. Philadelphia: University of Pennsylvania Press.

———. 2016. "Sensory Attunements: Working with the Past in the Little Cities of Black Diamonds." *South Atlantic Quarterly* 115(1): 89–111.

Picker, John M. 1999. "The Soundproof Study: Victorian Professionals, Work Space, and Urban Noise. *Victorian Studies* 42(3): 427–53.

Plourde, Lorraine. 2008. "Disciplined Listening in Tokyo: Onkyō and Non-Intentional Sounds." *Ethnomusicology* 52(2): 270–95.

———. 2009. "Difficult Music: An Ethnography of Listening for the Avant-Garde in Tokyo." PhD dissertation, Columbia University.

———. 2014. "Distraction, Noise, and Ambient Sounds in Tokyo." In *Sound, Space, and Sociality in Modern Japan*, edited by Joseph D. Hankins and Carolyn S. Stevens, 71–88. London: Routledge.

———. 2014. "Cat Cafes, Affective Labor, and the Healing Boom in Japan." *Japanese Studies* 34(2): 115–33.

———. 2017. "Sonic Air-Conditioning: Muzak as Affect Management for Office Workers in Japan." In *The Senses and Society* 12(1): 18–34.

Powers, Devon. 2010. "Strange Powers: The Branded Sensorium and the Intrigue of Musical Sound." In *Blowing Up the Brand: Critical Perspectives on Promotional Culture*, edited by Melissa Aronczyk and Devon Powers, 285–306. New York: Peter Lang.

Prochnik, George. 2010. *In Pursuit of Silence: Listening for Meaning in a World of Noise*. New York: Doubleday.

Radano, Ronald M. 1989. "Interpreting Muzak: Speculations on Musical Experience in Everyday Life." *American Music* 7(4): 448–60.

Rawson, Eric. 2006. Perfect Listening: Audiophilia, Ambiguity, and the Reduction of the Arbitrary. *Journal of American Culture* 29(2): 202–12.

Rice, Tom. 2015. "Listening." In *Keywords in Sound*, edited by David Novak and Matt Sakakeeny, 99–111. Durham: Duke University Press.

Roquet, Paul. 2008. "Ambient Literature and the Aesthetics of Calm: Mood Regulation in Contemporary Japanese Fiction." *Journal of Japanese Studies* 35(1): 87–111.

———. 2016. *Ambient Media: Japanese Atmospheres of Self*. Minneapolis: University of Minnesota Press.

Rothenbuhler, Eric W., and John Durham Peters. 1997. "Defining Phonography: An Experiment in Theory." *Musical Quarterly* 81(2): 242–64.

Russo, Mary, and Daniel Warner. 1987. "Rough Music, Futurism, and Postpunk Industrial Noise Bands." *Discourse* 10(1): 55–76.

Russolo, Luigi. 1986 [1916]. *The Art of Noises*. Translated by Barclay Brown. New York: Pendragon.

Sakamoto, Ryuichi, and Hosokawa Shuhei, eds. 1986. *Miraiha 2009* [Futurism 2009]. Tokyo: Honhondo.

Sand, Jordan. 2006. "The Ambivalence of the New Breed: Nostalgic Consumerism in 1980s and 1990s Japan." In *The Ambivalent Consumer: Questioning Consumption in East Asia and the West*, edited by Sheldon Garon and Patricia L. Maclachlan, 85–108. Ithaca: Cornell University Press.

———. 2008. "Street Observation Science and the Tokyo Economic Bubble, 1986–1990." In *The Spaces of the Modern City: Imaginaries, Politics and Everyday Life*, edited by Gyan Prakash and Kevin M. Krause, 373–400. Princeton: Princeton University Press.

———. 2013. *Tokyo Vernacular: Common Spaces, Local Histories, Found Objects*. Berkeley: University of California Press.

Sas, Miryam.1999. *Fault Lines: Cultural Memory and Japanese Surrealism*. Stanford: Stanford University Press.

———. 2005. "Subject, City, Machine." In *Histories of the Future*, edited by Daniel Rosenberg and Susan Harding, 203–23. Durham: Duke University Press.

Sasaki, Atsushi. 2001. "*Keiiji, minimarizumu, onkyō ha*" [Cage, Minimalism, *Onkyō ha*]. In *Tekunoizu Materiurizumu* [Technoise Materialism], 163–224. Tokyo: Seidosha.

———. 2005. *Tekuno/Rojikaru/Ongakuron: Shutokkuhauzen kara onkyō ha made* [Techno/Logical/Theory of Music: From Stockhausen to *Onkyō ha*]. Tokyo: Rittor Music.

———, ed. 2005. "*Tokushu: Posuto Noizu: ekkyō suru saundo*" [Special Issue, Post-Noise: Border-Crossing Sound]. *Yuriika* (Eureka) 37: 3. Tokyo: Seidosha.

Scannell, John. 2014. "'Do You Know the Way to San Jose?' Burt Bacharach and the Music of Non-Place." *Social Alternatives* 33(1): 73–79.

Schafer, R. Murray. 1977. *The Soundscape: Our Sonic Environment and the Tuning of the World*. Rochester, VT: Destiny.

Schrey, Dominik. 2014. "Analogue Nostalgia and the Aesthetics of Digital Remediation." In *Media and Nostalgia: Yearning for the Past, Present and Future*, edited by Katharina Niemeyer, 27–38. New York: Palgrave Macmillan.

Schull, Natasha Dow. 2012. *Addiction by Design: Machine Gambling in Las Vegas*. Princeton: Princeton University Press.

Seremetakis, C. Nadia. 1994. *The Senses Still: Perception and Memory as Material Culture in Modernity*. Chicago: University of Chicago Press.

Shattuck, Roger. 1955. *The Banquet Years: The Origins of the Avant-Garde in France, 1885 to World War I*. New York: Vintage.

Smith, Mark M. 2007. *Sensing the Past: Seeing, Hearing, Smelling, Tasting, and Touching in History*. Berkeley: University of California Press.

Sterne, Jonathan. 1997. "Sounds Like the Mall of America: Programmed Music and the Architectonics of Commercial Space." *Ethnomusicology* 4(1): 22–50.

———. 2003. *The Audible Past: Cultural Origins of Sound Production*. Durham: Duke University Press.

———. 2013. "The Non-Aggressive Music Deterrent." In *Ubiquitous Musics: The Everyday Sounds That We Don't Always Notice*, edited by Marta Garcia Quinones, Anahid Kassabian, and Elena Boschi, 121–37. Farnham, Surrey: Ashgate.

Stevens, Carolyn S. 2016. "Irasshai! Sonic Practice as Commercial Enterprise in Urban Japan." *Journal of Musicological Research* 35(2): 82–99.

Stewart, Kathleen. 2011. "Atmospheric Attunements." *Environment and Planning D: Society and Space* 29(3): 445–53.

Sumrell, Robert, and Kazys Varnelis. 2007. *Blue Monday: Stories of Absurd Realities and Natural Philosophies*. Barcelona: Actar.

Szendy, Peter. 2008. *Listen: A History of Our Ears*. Translated by Charlotte Mandell. New York: Fordham University Press.

Takahashi, Yasuo. 1994. "Why You Can't Have Green Tea in a Japanese Coffee Shop." In *The Electric Geisha: Exploring Japan's Popular Culture*, edited by Atsushi Ueda and translated by Miriam Eguchi, 26–33. Tokyo: Kodansha International.

Takeyama, Akiko. 2016. *Staged Seduction: Selling Dreams in a Tokyo Host Club*. Stanford: Stanford University Press.

Tanizaki, Jun'ichiro. 1977 [1933]. *In Praise of Shadows*. Sedgwick, ME: Leete's Island Books.

Taussig, Michael. 1991. "Tactility and Distraction." *Cultural Anthropology* 6(2):147–53.

Thompson, Emily. 2002. *The Soundscape of Modernity: Architectural Acoustics and the Culture of Listening in America, 1900–1933*. Cambridge, MA: MIT Press.

———. 2005. "Noise and Noise Abatement in the Modern City." In *Sense of the City: An Alternate Approach to Urbanism*, edited by Mirko Zardini, 190–99. Montreal: Lars Müller.

Thrift, Nigel. 2004. "Intensities of Feeling: Towards a Spatial Politics of Affect." *Geografiska Annaler* 86B(1): 57–78.

———. 2010. "Understanding the Material Practices of Glamour." In *The Affect Theory Reader*, edited by Melissa Gregg and Gregory J. Seigworth, 289–308. Durham: Duke University Press.

Tiekso, Tanja. 2016. "Luigi Russolo's Art of Noises and the Aesthetics of Musical Experimentalism." In *International Yearbook of Futurism Studies*, vol. 6, edited by Günter Berghaus, 71–93. Berlin: De Gruyter.

Tsing, Anna Lowenhaupt. 2015. *The Mushroom at the End of the World: On the Possibility of Life in Capitalist Ruins*. Princeton: Princeton University Press.

Tsunoda, Tadanobu. 1985. *The Japanese Brain: Uniqueness and Universality*. Translated by Y. Oiwa. Tokyo: Taishukan.

Vanel, Herve. 2008. "John Cage's Muzak-Plus: The Fu(rni)ture of Music." *Representations* 102: 94–128.

———. 2013. *Triple Entendre: Furniture Music, Muzak, Muzak-Plus*. Urbana: University of Illinois Press.

White, Merry. 2012. *Coffee Life in Japan*. Berkeley: University of California Press.

Yano, Christine R. 2002. *Tears of Longing: Nostalgia and the Nation in Japanese Popular Song*. Cambridge, MA: Harvard University Asia Center.

———. 2005. "Covering Disclosures: Practices of Intimacy, Hierarchy, and Authenticity in a Japanese Popular Music Genre." *Popular Music and Society* 28(2): 193–205.

Yasar, Kerim. 2009. "Electrified Voices: Media Technology and Discourse in Modern Japan." Doctoral dissertation, Columbia University.

Yoshimi, Shunya. 2011. "The Market of Ruins, or the Destruction of the Cultural City." *Japan Forum* 23(2): 287–300.

Yoshimura, Hiroshi. 1990. *Toshi no oto* [Sounds of the City]. Tokyo: Shunjusha.

Zardini, Mirko, ed. 2005. *Sense of the City: An Alternate Approach to Urbanism*. Montreal: Lars Müller.

INDEX

NOTE: Page numbers in *italics* indicate illustrations.

MUSIC / CULTURE

A series from Wesleyan University Press
Edited by Deborah Wong, Sherrie Tucker, and Jeremy Wallach
Originating editors: George Lipsitz, Susan McClary, and Robert Walser

Marié Abe
Resonances of Chindon-ya: Sounding Space and Sociality in Contemporary Japan

Frances Aparicio
Listening to Salsa: Gender, Latin Popular Music, and Puerto Rican Cultures

Paul Austerlitz
Jazz Consciousness: Music, Race, and Humanity

Harris M. Berger
Metal, Rock, and Jazz: Perception and the Phenomenology of Musical Experience

Harris M. Berger
Stance: Ideas about Emotion, Style, and Meaning for the Study of Expressive Culture

Harris M. Berger and Giovanna P. Del Negro
Identity and Everyday Life: Essays in the Study of Folklore, Music, and Popular Culture

Franya J. Berkman
Monument Eternal: The Music of Alice Coltrane

Dick Blau, Angeliki Vellou Keil, and Charles Keil
Bright Balkan Morning: Romani Lives and the Power of Music in Greek Macedonia

Susan Boynton and Roe-Min Kok, editors
Musical Childhoods and the Cultures of Youth

James Buhler, Caryl Flinn, and David Neumeyer, editors
Music and Cinema

Thomas Burkhalter, Kay Dickinson, and Benjamin J. Harbert, editors
The Arab Avant-Garde: Music, Politics, Modernity

Patrick Burkart
Music and Cyberliberties

Julia Byl
Antiphonal Histories: Resonant Pasts in the Toba Batak Musical Present

Daniel Cavicchi
Listening and Longing: Music Lovers in the Age of Barnum

Susan D. Crafts, Daniel Cavicchi, Charles Keil, and the Music in Daily Life Projec
My Music: Explorations of Music in Daily Life

Jim Cullen
Born in the USA: Bruce Springsteen and the American Tradition

Anne Danielsen*Presence and Pleasure: The Funk Grooves of James Brown and Parliament*

Peter Doyle
Echo and Reverb: Fabricating Space in Popular Music Recording, 1900–1960

Ron Emoff
Recollecting from the Past: Musical Practice and Spirit Possession on the East Coast of Madagascar

Yayoi Uno Everett and Frederick Lau, editors
Locating East Asia in Western Art Music

Susan Fast and Kip Pegley, editors
Music, Politics, and Violence

Heidi Feldman
Black Rhythms of Peru: Reviving African Musical Heritage in the Black Pacific

Kai Fikentscher
"You Better Work!" Underground Dance Music in New York City

Ruth Finnegan
The Hidden Musicians: Music-Making in an English Town

Daniel Fischlin and Ajay Heble, editors
The Other Side of Nowhere: Jazz, Improvisation, and Communities in Dialogue

Wendy Fonarow
Empire of Dirt: The Aesthetics and Rituals of British "Indie" Music

Murray Forman
The 'Hood Comes First: Race, Space, and Place in Rap and Hip-Hop

Lisa Gilman
My Music, My War: The Listening Habits of U.S. Troops in Iraq and Afghanistan

Paul D. Greene and Thomas Porcello, editors
Wired for Sound: Engineering and Technologies in Sonic Cultures

Tomie Hahn
Sensational Knowledge: Embodying Culture through Japanese Dance

Edward Herbst
Voices in Bali: Energies and Perceptions in Vocal Music and Dance Theater

Deborah Kapchan
Traveling Spirit Masters: Moroccan Gnawa Trance and Music in the Global Marketplace

Deborah Kapchan, editor
Theorizing Sound Writing

Max Katz
Lineage of Loss: Counternarratives of North Indian Music

Raymond Knapp
Symphonic Metamorphoses: Subjectivity and Alienation in Mahler's Re-Cycled Songs

Victoria Lindsay Levine and Dylan Robinson, editors
Music and Modernity among First Peoples of North America

Laura Lohman
Umm Kulthūm: Artistic Agency and the Shaping of an Arab Legend, 1967–2007

Preston Love
A Thousand Honey Creeks Later: My Life in Music from Basie to Motown—and Beyond

René T. A. Lysloff and Leslie C. Gay Jr., editors
Music and Technoculture

Ian MacMillen
Playing It Dangerously: Tambura Bands, Race, and Affective Block in Croatia and Its Intimates

Allan Marett
Songs, Dreamings, and Ghosts: The Wangga of North Australia

Ian Maxwell
Phat Beats, Dope Rhymes: Hip Hop Down Under Comin' Upper

Kristin A. McGee
Some Liked It Hot: Jazz Women in Film and Television, 1928–1959

Tracy McMullen
Haunthenticity: Musical Replay and the Fear of the Real

Rebecca S. Miller
Carriacou String Band Serenade: Performing Identity in the Eastern Caribbean

Tony Mitchell, editor
Global Noise: Rap and Hip-Hop Outside the USA

Christopher Moore and Philip Purvis, editors
Music & Camp

Rachel Mundy
Animal Musicalities: Birds, Beasts, and Evolutionary Listening

Keith Negus
Popular Music in Theory: An Introduction

Johnny Otis
Upside Your Head: Rhythm and Blues on Central Avenue

Kip Pegley
Coming to You Wherever You Are: MuchMusic, MTV, and Youth Identities

Jonathan Pieslak
Radicalism and Music: An Introduction to the Music Cultures of al-Qa'ida, Racist Skinheads, Christian-Affiliated Radicals, and Eco-Animal Rights Militants

Matthew Rahaim
Musicking Bodies: Gesture and Voice in Hindustani Music

John Richardson
Singing Archaeology: Philip Glass's Akhnaten

Tricia Rose
Black Noise: Rap Music and Black Culture in Contemporary America

David Rothenberg and
Marta Ulvaeus, editors
The Book of Music and Nature: An Anthology of Sounds, Words, Thoughts

Nichole Rustin-Paschal
The Kind of Man I Am: Jazzmasculinity and the World of Charles Mingus Jr.

Marta Elena Savigliano
Angora Matta: Fatal Acts of North-South Translation

Joseph G. Schloss
Making Beats: The Art of Sample-Based Hip-Hop

Barry Shank
Dissonant Identities: The Rock 'n' Roll Scene in Austin, Texas

Jonathan Holt Shannon
Among the Jasmine Trees: Music and Modernity in Contemporary Syria

Daniel B. Sharp
Between Nostalgia and Apocalypse: Popular Music and the Staging of Brazil

Helena Simonett
Banda: Mexican Musical Life across Borders

Mark Slobin
Subcultural Sounds: Micromusics of the West

Mark Slobin, editor
Global Soundtracks: Worlds of Film Music

Christopher Small
The Christopher Small Reader

Christopher Small
Music of the Common Tongue: Survival and Celebration in African American Music

Christopher Small
Music, Society, Education

Christopher Small
Musicking: The Meanings of Performing and Listening

Maria Sonevytsky
Wild Music: Sound and Sovereignty in Ukraine

Regina M. Sweeney
Singing Our Way to Victory: French Cultural Politics and Music during the Great War

Colin Symes
Setting the Record Straight: A Material History of Classical Recording

Steven Taylor
False Prophet: Field Notes from the Punk Underground

Paul Théberge
Any Sound You Can Imagine: Making Music/Consuming Technology

Sarah Thornton
Club Cultures: Music, Media, and Subcultural Capital

Michael E. Veal
Dub: Songscape and Shattered Songs in Jamaican Reggae

Michael E. Veal and
E. Tammy Kim, editors
Punk Ethnography: Artists and Scholars Listen to Sublime Frequencies

Robert Walser
Running with the Devil: Power, Gender, and Madness in Heavy Metal Music

Dennis Waring
Manufacturing the Muse: Estey Organs and Consumer Culture in Victorian America

Lise A. Waxer
The City of Musical Memory: Salsa, Record Grooves, and Popular Culture in Cali, Colombia

Mina Yang
Planet Beethoven: Classical Music at the Turn of the Millennium

ABOUT THE AUTHOR

Lorraine Plourde received her MA in ethnomusicology from the University of Washington and her PhD in sociocultural anthropology from Columbia University. Plourde's research has been funded by the Social Science Research Council (SSRC), Fulbright-Hays, and the Northeast Asia Council for the Association for Asian Studies. She has published articles on Japanese heavy metal pop idol girl group Babymetal, and Japan's cat café phenomenon. She is currently associate professor of media studies and anthropology at the State University of New York at Purchase.